INTERNATIONAL SERIES OF MONOGRAPHS IN
NATURAL PHILOSOPHY
GENERAL EDITOR: D. TER HAAR

VOLUME 77

THE PSEUDO-SPIN METHOD IN MAGNETISM AND FERROELECTRICITY

THE PSEUDO-SPIN METHOD IN MAGNETISM AND FERROELECTRICITY

By

LJUBISAV NOVAKOVIĆ, D.Phil. Oxford

Senior Research Associate of the Boris Kidrich Institute and Associate Professor of the Faculty of Mathematical and Natural Sciences, Belgrade

PERGAMON PRESS

OXFORD · NEW YORK · TORONTO
SYDNEY · BRAUNSCHWEIG

Pergamon Press Ltd., Headington Hill Hall, Oxford
Pergamon Press Inc., Maxwell House, Fairview Park, Elmsford, New York 10523
Pergamon of Canada Ltd., 207 Queen's Quay West, Toronto 1
Pergamon Press (Aust.) Pty. Ltd., 19a Boundary Street, Rushcutters Bay, N.S.W. 2011, Australia
Pergamon Press GmbH, Burgplatz 1, Braunschweig 3300, West Germany

---

First edition 1975

Library of Congress Cataloging in Publication Data

Novaković Ljubisav.

The speudo-spin method in magnetism and ferroelectricity.

(International series of monographs in natural philosophy, v. 77)

1. Ferromagnetism. 2. Ferroelectricity. 3. Nuclear spin. 4. Lattice dynamics. I. Title.

QC761.5.N68 1975 538'.44 74-17134

ISBN 0-08-018060-4

Printed in Hungary

Contents

Acknowledgements

I AM deeply indebted to Professor Sir Rudolf Peierls for a kind talk on several crucial points related to the problem of antiferromagnetism during my stay in Oxford in the winter 1972. I am also grateful to Professor W. Cochran and Dr. H. Montgomery for a warm hospitality as well as useful discussion on the present subject which I had during my stay in Edinburgh at the mentioned time. Also my thanks are due to Professor R. Blinc for critical comments upon reading the manuscript during my recent visit to the University of Ljubljana. Finally I wish to acknowledge the help and assistance to both Mr. F. Dacar of the Ljubljana University and Mr. A. Vlahov of the Boris Kidrich Institute for an important numerical work, Appendix D.

L. NOVAKOVIĆ

Introduction

ANY energy multiplet of a single atom subject to an external magnetic field is determined entirely by the spin angular momentum quantum number, S. On the other hand, a given energy multiplet as referred to the corresponding paramagnetic ions at particular lattice sites is determined entirely by the symmetry requirement as imposed on the lattice wave functions. In principle energy states of the multiplet are enumerated by different symmetry elements. Now if the lattice wave function is characterized by an abstract number to be named the pseudo-spin, S, then the total number of different symmetry elements, $2S+1$, will determine the total number of states in the energy multiplet. It is reasonable here to introduce the following physical assumption: all lattice states with the same value of the pseudo-spin are expected to have the same energy. Therefore the pseudo-spin concept although mathematically distinguished from the ordinary-spin concept will suffice to describe the dynamics of magnetism in just the same manner.

The present method may also be applied to the dynamical behavior of hydrogen-bonded ferroelectrics which happens to depend very considerably on the proton motion, the proton sites being apparently situated at two distinct equilibrium positions. Thus the latter concept is closely related to the pseudo-spin one-half. In general, whenever a group of atoms, or an individual atomic cluster, has two or more distinct positions so as to emphasize a neat energy difference between any two successive positions we may associate a certain pseudo-spin with them in order to classify the appropriate energy configurations. For this reason the hydrogen-bonded problem, being apparently similar to the Ising model with a transverse field, is reduced to a well-established mathematical method.

Once the basic interactions are established with the help of the pseudo-spin concept the problem then is reduced to a familiar method but it is still far from being solved exactly for all temperatures and for all ordered systems. To evaluate the order parameter, or the specific heat, or any other characteristic quantity one has to transform the pseudo-spin variables into a set of auxiliary creation and annihilation operators with a defined statistical behavior. Three important ordered systems are considered in the present work, ferromagnets, antiferromagnets, and hydrogen-bonded ferroelectrics, each of them in three important temperature regions: low temperatures, the critical region, as well as high temperatures. The presented mathematical method is predominantly analytical in the sense that all crucial physical quantities are expressed in terms of temperature or the reciprocal lattice vectors.

CHAPTER 1

Lattice Dynamics

1.1. Basic concepts in symmetry groups

By a solid body we understand the substance which has a certain stiffness under pressure. Such substances are normally associated with crystal structures, each of them being described by a type of the lattice and a certain structure motive. The crystal lattice consists of atomic sites arranged in space in such a way that they enable the introduction of three vectors of elementary translations $\boldsymbol{a}$, $\boldsymbol{b}$, and $\boldsymbol{c}$ with the following property. The crystal lattice viewed from the given point $\boldsymbol{R}$ has the same shape as viewed from another given point $\boldsymbol{R}'$ such that

$$\boldsymbol{R}' = \boldsymbol{R}+n_1\boldsymbol{a}+n_2\boldsymbol{b}+n_3\boldsymbol{c} \tag{1.1}$$

where n_j designate three integers (positive, negative, or zero). The basic vectors of elementary translations are often selected to coincide with the crystallographic axes. The structure motive consists of atomic groups, each atom being associated with one lattice site. These atomic groups are identical with respect to the content, arrangement, and orientation. The simplest possible case occurs when the atomic groups contain only the single atoms, all being of the same kind.

A set of all symmetry transformations (operations) of a given body is called the symmetry group. The symmetry of a body is realized by a set of translations which bring the body into itself. Each of these translations can be decomposed into a combination of three basic transformations:

1. a rotation with respect to an axis;
2. the mirror reflection with respect to a plane; and
3. a parallel displacement of the body for a certain distance.

A more detailed exposition can be found in the monographs by Kittel,(1) Levy,(2) Landau and Lifshitz,(3) and Lyubarskii.(4)

The former two transformations apply to a body with a finite size, such as a molecule, whereas the third transformation applies only to an infinite means such as the crystal lattice.

Every transformation which brings a vector $\boldsymbol{R}$ into $-\boldsymbol{R}$ is an inversion. All rotations form a group named the rotation group; if here the inversion is included then there appears the complete orthogonal group. Any subgroup of the complete orthogonal group is called the point group. There are thirty-two possible point groups for crystals but we shall mention explicitly only three, namely $\mathcal{G}_n(n)$, $\mathcal{D}_{2d}(\bar{4}2m)$ and $\mathcal{O}(432)$ as follows.

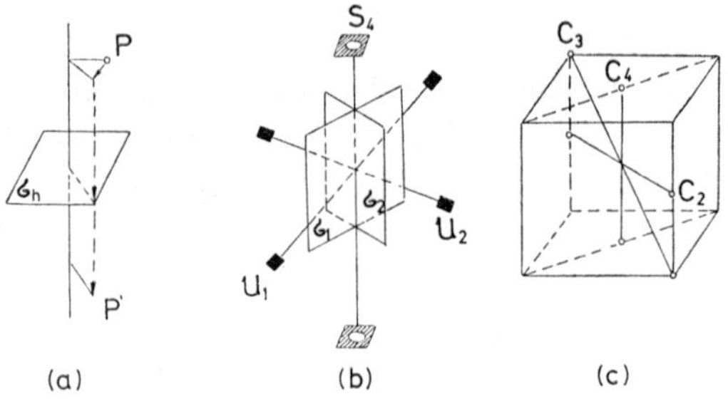

FIG. 1.1. An illustration for the space point groups: (a) $\mathcal{G}_n(n)$; (b) $\mathcal{D}_{2d}(\bar{4}2m)$, and (c) $\mathcal{O}(432)$.

1. The space point group $\mathcal{G}_n$ contains all rotations with respect to an nth-order axis, the rotations being defined by the angles $2k\pi/n$ where $k = 1, 2, \ldots, n$. It is a cyclic group and contains n elements. Since every element is a class by itself it appears that $\mathcal{G}_n$ has n classes (see Fig. 1.1).

2. The space point group $\mathcal{D}_{2d}$ is a special case of the point group $\mathcal{D}_{nd}$ which is the symmetry group of the following body. Let us imagine two identical regular prisms, each having the basis in the form of a regular n-sided polygon, and the prisms being placed one on the top of the other and rotated one with respect to the other through an angle π/n. When viewed from the top the two bases look like a regular $2n$-sided polygon. Now the point group $\mathcal{D}_{nd}$ contains one $2n$th-order axis of mirror reflections, n vertical planes passing through the edges of a given prism, and n horizontal second-order axes passing through the intersections of the two regular n-sided polygons. For $n = 2$ we obtain one fourth-order axis of mirror reflections, two vertical planes σ_1 and σ_2 being placed normally one to another, and two horizontal axes u_1 and u_2 being also placed normally one to another, but at the same time having bisected the angle closed by the vertical planes (see Fig. 1.1). The point group $\mathcal{D}_{2d}$ has eight elements; these are classified into the following five classes:

one class coming from all mirror reflections with respect to the vertical planes;
another class coming from all rotations with respect to the horizontal axes; and
another three classes coming from a successive application of the rotations with respect to the two-sided axis S_4,

$$\{E\}, \quad \{S_4, S_4^3\}, \quad \{S_4^2\},$$

where E designates the unit element.

3. The space point group $\mathcal{O}$ contains all rotations which bring a cube or hexahedron into itself (see Fig. 1.1). This group contains twenty-four elements classified into five classes.

One of basic concepts bridging the group theory with a variety of applications to spectroscopy is the group representation which is known to be defined as a certain association of the set of linear operators with the group elements. By definition, a product of any two group elements is associated with the product of the linear operators of the corresponding elements. We say that the set of such linear operators form a representation of the given group. Any representation can be expressed in a matrix form. Using a similarity transformation some of the above representations can be reduced to a diagonal form; these are named the irreducible representations.

The sum over the diagonal matrix elements realizing a given representation is named the character of the representation. It is clear that the character of a given representation is invariant under a change of the basis in the space of linear operators. As we know the group elements can be divided into classes, and all elements belonging to a given class have one and the same character. In addition, the number of irreducible representations is equal to the number of classes of a given group. To illustrate the present subject we bring the characters for the space point groups $\mathcal{D}_{2d}$ and $\mathcal{O}$ which belong to the tetragonal and cubic systems, respectively (see Tables 1.1 and 1.2). In addition,

TABLE 1.1.

Characters for the Space Point Group $\mathcal{D}_{2d}$ ($\bar{4}2m$)

	E	$2S_4$	C_2	$2C_2'$	$2\sigma_d = \sigma_1+\sigma_2$
A_1	1	1	1	1	1
A_2	1	1	1	-1	-1
B_1	1	-1	1	1	-1
B_2	1	-1	1	-1	1
E	2	0	-2	0	0

TABLE 1.2.

Characters for the Space Point Group $\mathcal{O}$ (432)

	E	$8C_3$	$3C_2$	$6C_4$	$6C_2'$
A_1	1	1	1	1	1
A_2	1	1	1	-1	-1
E	2	-1	2	0	0
F_1	3	0	-1	1	-1
F_2	3	0	-1	-1	1

the number of irreducible representations is equal to the number of classes of a given group. Except in the cited references 3–4 the present method is also developed by Wilson, Decius and Cross,[5] and Tinkham.[6]

The one-dimensional representations are designated by the letters A and B, the two-dimensional representations by E, and the three-dimensional representations by F, as indicated in the left column in Tables 1.1 and 1.2. Here A and B are used to distinguish between the symmetric and antisymmetric representations, respectively, with respect to the generating operation C_n in the space point groups $\mathcal{D}_n$. Numerical subscripts are employed with A or B to designate the case of degenerate representations with respect to one of the two-fold rotations about an axis perpendicular to the principal symmetry axis in $\mathcal{D}_n$. A similar definition holds for the numerical subscripts with F. In both cases the z-axis is selected to be either parallel or identical with the principal symmetry axis. The numbers standing in front of the symbols which designate the group elements in the top row indicate how many group elements belong to the corresponding classes. A total sum of those numbers is equal to the number of group elements.

To illustrate the above introduced concepts we consider the space point group $\mathcal{D}_{2d}$ explicitly. Here the group elements are identified with the rotations through the angle $\varphi = k\pi/2$; in particular, the rotations are S_4, $C_2 \equiv S_4^2$, S_4^3, and $E \equiv S_4^4$ which correspond to $k = 1, 2, 3,$ and 4. Having written the two-dimensional representations as

$$E \sim \begin{bmatrix} \exp(i\varphi) & 0 \\ 0 & \exp(-i\varphi) \end{bmatrix}$$

we obtain at once

	S_4	C_2	S_4^3	E	σ_1
$E \sim$	$\begin{bmatrix} i & 0 \\ 0 & -i \end{bmatrix}$	$\begin{bmatrix} -1 & 0 \\ 0 & -1 \end{bmatrix}$	$\begin{bmatrix} -i & 0 \\ 0 & i \end{bmatrix}$	$\begin{bmatrix} 1 & 0 \\ 0 & 1 \end{bmatrix}$	$\begin{bmatrix} 0 & 1 \\ 1 & 0 \end{bmatrix}$

(1.2)

Any other representation can be obtained with the help of the following multiplication law. If, for instance,

$$u_1 = S_4\sigma_1, \quad u_2 = S_4\sigma_2$$

then there follows

$$E(u_1) = E(S_4)\,E(\sigma_1),$$
$$E(u_2) = E(S_4)\,E(\sigma_2).$$

Also having taken the one-dimensional representations as

$$\begin{array}{llllll} A_1 & \text{and} & A_2 & \text{of} & S_4 \sim 1, \\ B_1 & \text{and} & B_2 & \text{of} & S_4 \sim -1, \\ A_1 & \text{and} & B_2 & \text{of} & \sigma_1 \sim 1, \\ A_2 & \text{and} & B_1 & \text{of} & \sigma_1 \sim -1, \end{array}$$

we have evaluated the characters for the space point group $\mathcal{D}_{2d}$ and presented them in Table 1.1. A similar method applies to the characters for the space point group $\mathcal{O}$.

1.2. Light scattering

The absorption or emission spectrum which arises from the rotational motion of a molecule having an electric moment is predominantly in the infrared region which covers the frequencies below approximately 200 wave numbers.† The absorption or emission spectrum arising from the vibrational motion of the same molecule is in the energy region from 200 up to about 3500 wave numbers. If the molecule is excited by monoenergetic light coming from a powerful laser then the spectrum consists of a strong line of the same frequency as the incident laser illumination as well as of some weaker lines on both sides with

† Energy can be expressed in units of the wave numbers or the reciprocal centimeter, cm^{-1}. A relationship to the usual energy units is given in Appendix E.

respect to the central line. Actually the lines on the low-frequency side are much more intense (Stokes lines) than those on the high-frequency side (anti-Stokes lines). Common name for them all is the Raman lines.

An explanation of the Raman effect is provided by quantum mechanics. As we know, any motion of the molecule is restricted to a set of discrete rotational or vibrational energy levels corresponding to certain stationary states. A transition from one energy level, E_i, to another, E_f, which the molecule undergoes under the influence of incident light is accompanied by the emission or absorption of light quanta whose frequency is ν_{fi}. The energy difference is given by

$$h\nu_{fi} = E_i - E_f$$

where h is Planck's constant.

Not all transitions can occur with the absorption or emission of radiation but only those obeying certain selection rules. According to them some transitions are allowed the others are forbidden. We can visualize the Raman effect by thinking of a photon which comes up to the molecule in a given stationary state and causes an allowed transition to another stationary state. If the energy of the photon is $h\nu_0$, ν_0 being the frequency of the incident light, then the energy of the scattered photon is $h(\nu_0 \pm \nu_{fi})$, the upper and lower signs correspond to the molecule in the higher and lower energy levels, respectively. In general more molecules are in the lower energy levels, since the photon has lost some of its energy, so the Stokes lines are more intense than the anti-Stokes lines.

In order to determine the selection rules one has to remember that the complete wave function corresponding to the given stationary state of the molecule can be written in a factorized form

$$\psi = \psi_V \psi_R \psi_T \tag{1.3}$$

where the letters used designate the vibrational, rotational and translational wave functions, respectively. (It is possible to show that the electronic wave function does not enter the problems treated here owing to the fact that no electronic excitation of the molecule is induced.) The molecule is said to be polarized by an electric field, $\boldsymbol{E}$, when the displacement of the charges caused by the electric field produce or alter the electric moment of the molecule, $\boldsymbol{\mu}$. If α designates the polarizability for an isotropic molecule then

$$\mu = \alpha \boldsymbol{E}. \tag{1.4}$$

Here the vector μ has the components

$$\mu_x = \sum_j e_j X_j,$$
$$\mu_y = \sum_j e_j Y_j,$$
$$\mu_z = \sum_j e_j Z_j,$$

where e_j designates the charge and X_j, Y_j, Z_j are cartesian coordinates with respect to the space-fixed axes for the jth molecule, the sum being taken over all the molecules. In general, however, the polarizability (1.4) for an anisotropic molecule must be replaced by a tensor entity,

$$\mathcal{A} = \begin{bmatrix} \alpha_{xx} & \alpha_{xy} & \alpha_{xz} \\ \alpha_{yx} & \alpha_{yy} & \alpha_{yz} \\ \alpha_{zx} & \alpha_{zy} & \alpha_{zz} \end{bmatrix} \tag{1.5}$$

where the introduced quantities $\alpha_{FF'}$ are independent of the components of the electric field. However, they depend on the orientation of the molecule with respect to the space-fixed axes. So the expression (1.4) is replaced by

$$\mu_x = \alpha_{xx} E_x + \alpha_{xy} E_y + \alpha_{xz} E_z,$$
$$\mu_y = \alpha_{yx} E_x + \alpha_{yy} E_y + \alpha_{yz} E_z,$$
$$\mu_z = \alpha_{zx} E_x + \alpha_{zy} E_y + \alpha_{zz} E_z.$$

It will be more appropriate in what follows to work with a rotating coordinate system of axes x, y, z rather than the space-fixed ones. The two systems of axes are connected one with another by certain linear relations.

Now it is easy to prove that the transformation properties of the electric moment, $\boldsymbol{\mu}$, are identical with those of the relative position vector, $\boldsymbol{r}$, in the sense that μ_x, μ_y, μ_z transform like x, y, z respectively. A similar proof holds for the transformation properties for the polarizability components; namely, the components α_{xx}, α_{xy} and so on transform like the bilinear forms xx, xy, and so on. The presented method is that of reference 5. It is immediately clear on the basis of the transformation properties that the polarizability is a symmetric tensor.

1.3. Selection rules for the translation and tensor representations

In classifying the crystal fundamentals one need to consider only the nonequivalent points which are contained in a three-dimensional Bravais unit cell. As we know, there are three pure translations or acoustic modes and $(3n-3)$ optical modes, n being the total number of nonequivalent points. The optical vibrations might be either external or internal depending on the forces holding the molecules.

If one assumes that the crystal is composed of identical structure clusters which are endlessly repeated in space then we expect that the forces holding the molecules together within the cluster are much stronger than those binding various clusters together in the crystal. The internal vibrations are those which arise predominantly from the motion of the molecules within the same cluster. The external vibrations however which are also named the lattice vibrations come from the motion of the clusters one relative to another. Owing to our assumption that the forces acting among various clusters are much too weak in comparison with those acting between the molecules within the same cluster one should expect that external vibrations will occur at frequencies too much low in comparison with those coming from the internal vibrations. The lattice vibrations are usually divided into rotational and translational modes which correspond to pure rotations and translations, respectively, in the limiting case where the forces acting among various clusters vanish.

In order to present the infrared and Raman activities associated with the normal lattice modes one can readily use group theory. The problem is then reduced to the evaluation of the selection rules respectively for the vector (translation) and symmetric-tensor operations by employing the following theorem: The character of an arbitrary reducible representation can be expanded in terms of the characters of the irreducible representations as proved by Lyubarskii[(4)] and Smirnov.[(7)]

Let g, q, and n designate respectively the group element, the total number of irreducible representations (which is equal to the total number of classes) and the total number of group elements; then the above theorem can be written

$$\chi(g) = \sum_{\alpha=1}^{q} C_\alpha \chi_\alpha(g) \tag{1.6}$$

where the expansion coefficients, C_α, are determined from the condition

$$C_\alpha = \frac{1}{n} \sum_g \chi_\alpha^*(g)\, \chi(g), \tag{1.7a}$$

the above sum being taken over all group elements. (As we know, all elements which belong to a given class have one and the same character.) Let β, g_β, and h_β designate respectively a given class, a group element which belongs to this class and the total number of class elements, then equation (1.7a) can be written in another form,

$$C_\alpha = \frac{1}{n} \sum_{\beta=1}^{q} h_\beta \chi_\alpha^*(g_\beta)\, \chi(g_\beta), \tag{1.7b}$$

the above sum being taken over all classes.

For a given vector (translation) operation, T, involving a rotation, R, about the z-axis (with + denoting a proper and − an improper rotation†), the character of the transformation of the translation com-

† By definition the improper rotation is the proper one followed by a reflection in a plane normal to the rotation axis.

ponents is evaluated as

$$\chi(T) = \pm 1 + 2\cos\phi_R. \tag{1.8}$$

Also for a given symmetric-tensor operation, $\mathcal{A}$, involving the same rotation, the character of the transformation of the tensor coordinates may be evaluated as

$$\chi(\mathcal{A}) = 2\cos\phi_R(\pm 1 + 2\cos\phi_R). \tag{1.9}$$

The selection rules for the infrared active fundamentals, that is the electric dipole moment vector, are given by

$$\frac{1}{n}\sum_{\beta=1}^{q} h_\beta \chi_\alpha^*(g_\beta)\,\chi(T)\begin{cases} = 0 \\ \neq 0 \end{cases}, \tag{1.10}$$

the two values corresponding to the infrared forbidden and allowed transitions, respectively. Also the selection rules for the Raman active fundamentals, that is the polarizability tensor, are given by

$$\frac{1}{n}\sum_{\beta=1}^{q} h_\beta \chi_\alpha^*(g_\beta)\,\chi(\mathcal{A})\begin{cases} = 0 \\ \neq 0 \end{cases}, \tag{1.11}$$

the two values corresponding to the Raman forbidden and allowed transitions, respectively. The numerical result is presented in Table 1.3 and Table 1.4 for the space point groups $\mathcal{D}_{2d}$ and $\mathcal{O}$, respectively.

TABLE 1.3.

Selection Rules for the Infrared and Raman Activities of the Space Point Group $\mathcal{D}_{2d}$ $(\bar{4}2m)$

Representation	Infrared	Raman
$A_1\ (x^2+y^2,\ z^2)$	forbidden	allowed
A_2	forbidden	forbidden
$B_1\ (x^2-y^2)$	forbidden	allowed
$B_2\ (xy;\ z)$	allowed	allowed
$E\ (xz,\ yz;\ x,\ y)$	allowed	allowed

Note. The following rotation angles are taken: $\phi_R = 0, \pi/2, \pi, \pi$ and 0 corresponding to the symmetry operations E, S_4, C_2, C_2', and σ_d, respectively.

TABLE 1.4.

Selection Rules for the Infrared and Raman Activities of the Space Point Group O (432)

Representation	Infrared	Raman
$A_1\ (x^2+y^2+z^2)$	forbidden	allowed
A_2	forbidden	forbidden
$E\ (x^2+y^2-2z^2,\ x^2-y^2)$	forbidden	allowed
$F_1\ (x,\ y,\ z)$	allowed	forbidden
$F_2\ (xy,\ yz,\ zx)$	forbidden	allowed

Note. The following rotation angles are taken: $\phi_R = 0,\ 2\pi/3,\ \pi,\ \pi/2$ and π corresponding to the symmetry operations E, C_3, C_2, C_4, and C_2', respectively.

The obtained result for the selection rules is in agreement with that of Mitra[(8)], Loudon[(9)], and the cited references 5–6. The derived selection rules will prove a great help in studying observed spectra of ferroelectric crystals. What is more, on the cubic point group, O, they will prove valuable in calculating the relevant vibrational optical modes for a number of ionic lattices which seem to provide an idealized theoretical model for lattice dynamics in hydrogen-bonded ferroelectrics at least.

1.4. The reciprocal lattice

All physical quantities which characterize the crystal lattice have the same periodicity as the lattice itself (the electromagnetic field induced by atoms, or the charge density taken over the ions, and so on). Let one of those quantities be $U(\boldsymbol{R})$ (with $\boldsymbol{R}$ being the position vector), so that

$$U(\boldsymbol{R}+n_1\boldsymbol{a}+n_2\boldsymbol{b}+n_3\boldsymbol{c}) = U(\boldsymbol{R}) \tag{1.12}$$

Now the periodic function U can be expanded in a three-fold Fourier series as follows:

$$U(\boldsymbol{R}) = \sum_{\boldsymbol{G}} U(\boldsymbol{G}) \exp(i\boldsymbol{G}\cdot\boldsymbol{R}) \tag{1.13}$$

where the sum is taken over all allowed values of the vector $\boldsymbol{G}$. The allowed values are determined by demanding that the function U satisfies the periodicity condition (1.12). The necessary and sufficient condition to meet this demand is that the scalar product appearing in the exponential factor be an integer multiple of 2π (positive, negative, or zero).

Writing

$$\begin{aligned} \boldsymbol{a}\cdot\boldsymbol{G} &= 2\pi p_1, \\ \boldsymbol{b}\cdot\boldsymbol{G} &= 2\pi p_2, \\ \boldsymbol{c}\cdot\boldsymbol{G} &= 2\pi p_3 \end{aligned} \tag{1.14}$$

we obtain the general solution for $\boldsymbol{G}$ as follows:

$$\boldsymbol{G} = 2\pi(p_1\boldsymbol{a}^* + p_2\boldsymbol{b}^* + p_3\boldsymbol{c}^*) \tag{1.15}$$

where the introduced vectors each marked by an asterisk are named the basic reciprocal lattice vectors of elementary translations. By definition

$$\begin{aligned} \boldsymbol{a}^* &= \frac{\boldsymbol{b}\times\boldsymbol{c}}{v}, \\ \boldsymbol{b}^* &= \frac{\boldsymbol{c}\times\boldsymbol{a}}{v}, \\ \boldsymbol{c}^* &= \frac{\boldsymbol{a}\times\boldsymbol{b}}{v}, \\ v &= [\boldsymbol{a}\cdot(\boldsymbol{b}\times\boldsymbol{c})], \end{aligned} \tag{1.16}$$

v being the volume of a unit cell. There follow mutual orthogonality relations between the basic vectors of the direct lattice space and those of the reciprocal lattice space ($\boldsymbol{a}^*$ is normal to $\boldsymbol{b}$ and $\boldsymbol{c}$, $\boldsymbol{b}^*$ is normal to $\boldsymbol{c}$ and $\boldsymbol{a}$, and so on). This enables one to conclude that the volume of a unit cell in the reciprocal lattice space is equal to the reciprocal value of the volume of unit cell in the direct lattice space. Indeed, by definition

$$v^* = [\boldsymbol{a}^* \cdot (\boldsymbol{b}^* \times \boldsymbol{c}^*)]$$

which after having substituted equations (1.16) becomes

$$v^* = \frac{1}{v^3} [\boldsymbol{b} \cdot (\boldsymbol{c} \times \boldsymbol{a})][\boldsymbol{c} \cdot (\boldsymbol{a} \times \boldsymbol{b})] = \frac{1}{v} .$$

It is convenient for a future analysis to introduce the reciprocal lattice vector with a slightly different form. In essence all written definitions will remain correct except that the vector $\boldsymbol{G}$ is replaced by another vector, $\boldsymbol{q}$, whose components are determined as follows. Let the crystal have N_1, N_2, and N_3 unit cells along the three basic directions, so that the total number of unit cells be equal to $N = N_1 N_2 N_3$; then the new reciprocal lattice vector is given by

$$\boldsymbol{q} = 2\pi \left(\frac{\nu_1}{N_1} \boldsymbol{a}^* + \frac{\nu_2}{N_2} \boldsymbol{b}^* + \frac{\nu_3}{N_3} \boldsymbol{c}^* \right) . \tag{1.17}$$

Here $\boldsymbol{q}$ has exactly N different values with the integers ν_j satisfying the relations

$$-\frac{N_j}{2} \leqslant \nu_j \leqslant \frac{N_j}{2} , \tag{1.18}$$

$j = 1, 2, 3$. (Each ν_j takes exactly N_j different values.) The quantity $\boldsymbol{q}$ is named the phonon wave vector; the reciprocal of q multiplied by 2π,

$$\lambda = \frac{2\pi}{q} ,$$

with

$$q = \sqrt{(q_x^2 + q_y^2 + q_z^2)} \tag{1.19}$$

is named the phonon wavelength.

The vectors (1.17) form the first Brillouin zone; this is a cell in the reciprocal lattice space whose points are closer to the origin than the points in another cell of the reciprocal lattice. The first three Brillouin zones for a two-dimensional lattice are presented in Fig. 1.2. It is

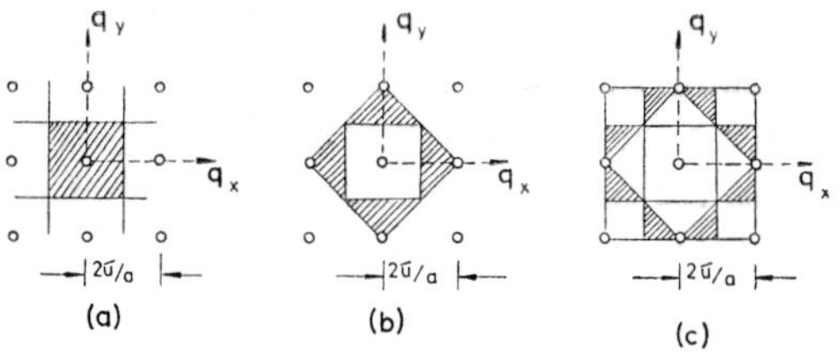

FIG. 1.2. An illustration for the first, second, and third Brillouin zones in a two-dimensional reciprocal lattice space.

easy to prove that the vectors $\boldsymbol{q}$ and $\boldsymbol{q}+\boldsymbol{G}$, where $\boldsymbol{G}$ designates any reciprocal lattice vector, are physically equivalent.

In specifying the position and orientation of the crystalline planes we use the well-known Miller indices, (hkl); they are determined as follows:

1. find the points where a given plane intersects the basic coordinate axes by writing the intersections in terms of the lattice constant; and
2. Take the reciprocal value of the obtained numbers by reducing them to the least common multiple, (hkl).

To label the directions in a lattice space we use the indices which are proportional to the components of a vector parallel to the given direction in the same coordinate system, $[hkl]$. It is easy to prove that the direction $[hkl]$ specified for all orthorhombic, tetragonal and cubic Bravais lattices is normal to the plane (hkl).

1.5. Basic concepts in lattice dynamics

Atoms or molecules will not sit still in their equilibrium positions even in the ideal crystal lattice owing to various reasons. The chain of masses connected by springs can be set into vibration by various forces (heat, incident acoustic waves, electromagnetic radiation, neutron scattering). The dispersion law for the lattice waves, named phonons, can be calculated exactly only for a limited number of simple physical situations. We shall expose here the method of obtaining the dispersion law for a three-dimensional diatomic crystal system which can be applied to the sodium-chloride structure, NaCl, with the following assumptions:

1. adiabatic approximation;
2. nearest-neighbor interactions;
3. harmonic expansion for the potential energy; and
4. no local fields present.

The fourth assumption, however, is essential since any account of the local fields affects the transverse and longitudinal modes in a different way.

The hamiltonian of the system consists of the kinetic and potential energies due to the motion of heavy ions with respect to the fixed equilibrium positions,

$$\mathcal{H} = T+U,$$
$$T = \tfrac{1}{2}\sum_{f=1}^{N}\sum_{\alpha=1}^{n} M_{\alpha}^{-1}p_{f\alpha}^{2},$$
$$U = U_0+\tfrac{1}{2}\sum_{f,g=1}^{N}\sum_{\alpha,\beta=1}^{n} K(fg;\alpha\beta)u_{f\alpha}u_{g\beta} \qquad (1.20)$$

where (f, g) and (α, β) designate respectively the unit cells and the heavy ions of mass M_α within the unit cells including the three space coordinates; in particular, an arbitrary diatomic unit cell ($n = 2$) is specified by

$$\alpha \equiv (f;\, x_1, y_1, z_1;\, x_2, y_2, z_2),$$
$$\beta \equiv (g;\, x_1, y_1, z_1;\, x_2, y_2, z_2).$$

The expansion coefficients $K(fg; \alpha\beta)$ are named the spring or elastic force constants. According to the second above listed assumption the nonvanishing values are

$$K(fg;\alpha\beta)^{\dagger} = \begin{cases} -K & \text{for nearest neighbors,} \\ 2K & \text{if } f = g, \quad \alpha = \beta. \end{cases} \qquad (1.21)$$

† In general the elastic force constants obey only the Ziman's cosmological principle[10] by which every $K(fg;\alpha\beta)$ depends only on the relative distance $|\boldsymbol{R}_g - \boldsymbol{R}_f|$.

The introduced displacement and momentum operators obey the quantum-mechanical commutation relation,

$$[u_{f\alpha}, p_{g\beta}] = i\hbar\, \Delta(f-g)\, \Delta(\alpha-\beta). \tag{1.22}$$

This relation has an actual influence only on the identical atoms within the same unit cell. It is clear that a Fourier transform leads to a similar relationship,

$$u_{f\alpha} = N^{-1/2} \sum_{\boldsymbol{q}} u_{\boldsymbol{q}\alpha} \exp(-i\boldsymbol{q}\cdot\boldsymbol{R}_f),$$

$$p_{f\alpha} = N^{-1/2} \sum_{\boldsymbol{q}} p_{\boldsymbol{q}\alpha} \exp(i\boldsymbol{q}\cdot\boldsymbol{R}_f),$$

$$[u_{\boldsymbol{q}\alpha}, p_{\boldsymbol{q}'\alpha'}] = i\hbar\, \Delta(\boldsymbol{q}-\boldsymbol{q}')\, \Delta(\alpha-\alpha'). \tag{1.23}$$

Here $\boldsymbol{R}_f$ designates the position vector referring to the center of a given unit cell, the above sums are taken over the first Brillouin zone. It is easy to see that the transformed displacement and momentum operators are hermitian conjugate, as follows:

$$u_{\boldsymbol{q}\alpha} = u^*_{-\boldsymbol{q}\alpha},$$

$$p_{\boldsymbol{q}\alpha} = p^*_{-\boldsymbol{q}\alpha}.$$

Using the above equations the hamiltonian (1.20) is readily transformed into[†]

$$\mathscr{H} = \tfrac{1}{2} \sum_{\boldsymbol{q}} \left\{ \sum_{\alpha=1}^{n} M_\alpha^{-1} p^*_{\boldsymbol{q}\alpha} p_{\boldsymbol{q}\alpha} + \sum_{\alpha,\beta=1}^{n} K_{\alpha\beta}(\boldsymbol{q})\, u^*_{\boldsymbol{q}\alpha} u_{\boldsymbol{q}\beta} \right\} \tag{1.24}$$

with

$$K_{\alpha\beta}(\boldsymbol{q}) = \sum_{f,g} K(fg;\alpha\beta) \exp[i\boldsymbol{q}(\boldsymbol{R}_g-\boldsymbol{R}_f)] = K^*_{\alpha\beta}(\boldsymbol{q}) = K_{\beta\alpha}(-\boldsymbol{q}).$$

Frequencies of the normal atomic transverse vibrations are determined by solving the equation of motion

$$M_\alpha \ddot{u}_{\boldsymbol{q}\alpha} = -\sum_{\beta=1}^{n} K_{\alpha\beta}(\boldsymbol{q}) u_{\boldsymbol{q}\beta} \tag{1.25}$$

† Here and elsewhere the reciprocal lattice vector is placed as suffix when designating an operator but as independent variable when designating a function.

for all values of α. The solution to this equation can be written

$$u_{q\alpha}(t) = M_{\alpha}^{-1/2} e_{\alpha}(\boldsymbol{q}) \exp(i\omega_T t) \tag{1.26}$$

with t designating the local time. Quantities $e_{\alpha}(\boldsymbol{q})$ are the well-known polarization vectors. When the above solution is inserted into equation (1.25) there follows the system of linear equations

$$\sum_{\beta=1}^{n} \{(M_{\alpha}M_{\beta})^{-1/2} K_{\alpha\beta}(\boldsymbol{q}) - \omega_T^2 \Delta(\alpha-\beta)\}\, e_{\beta}(\boldsymbol{q}) = 0 \tag{1.27a}$$

or equivalently

$$\det |(M_{\alpha}M_{\beta})^{-1/2} K_{\alpha\beta}(\boldsymbol{q}) - \omega_T^2 \Delta(\alpha-\beta)| = 0. \tag{1.27b}$$

In general there are three acoustic branches (vibrations) whose frequency is characterized by

$$\lim \omega_T = 0 \tag{1.28}$$

as $\boldsymbol{q} \to 0$. The remaining $(3n-3)$ solutions ($n = 2$ for diatomic unit cells) are optical branches (vibrations) whose frequencies do not vanish as $\boldsymbol{q} \to 0$.

1.6. The spectrum of ionic crystals

In ionic crystals the electrons make frequent transitions from the atom of one type to the atom of another type, so the crystal looks all the time as if it were composed of positive and negative ions separately. The ions are distributed in such a way that an electrostatic attraction between the ions of opposite sign is stronger than the repulsion between the identical ions. A degree of ionization of the atoms entering the ionic crystal is so high that the electron shells of the constituent atoms resemble those of the neighbor noble-gas atoms. This may be illustrated by looking at the sodium-chloride structure.

Here the free ions are represented by the following electron configurations

$$_{11}\mathrm{Na}^{(+)}: \quad 1s^2\, 2s^2\, 2p^6,$$
$$_{17}\mathrm{Cl}^{(-)}: \quad 1s^2\, 2s^2\, 2p^6\, 3s^2\, 3p^6,$$

which are identical with those of neon and argon, respectively. This ionic crystal has a face-centered cubic structure where each sodium ion is surrounded by six nearest chlorine ions, and by twelve next-nearest sodium ions.

Using the method exposed in the previous section frequencies of the normal atomic transverse vibrations are given by

$$\begin{vmatrix} A_1-\omega_T^2 & 0 & 0 & B(\boldsymbol{q}) & 0 & 0 \\ 0 & A_1-\omega_T^2 & 0 & 0 & B(\boldsymbol{q}) & 0 \\ 0 & 0 & A_1-\omega_T^2 & 0 & 0 & B(\boldsymbol{q}) \\ B(\boldsymbol{q}) & 0 & 0 & A_2-\omega_T^2 & 0 & 0 \\ 0 & B(\boldsymbol{q}) & 0 & 0 & A_2-\omega_T^2 & 0 \\ 0 & 0 & B(\boldsymbol{q}) & 0 & 0 & A_2-\omega_T^2 \end{vmatrix} = 0 \quad (1.29)$$

where

$$A_1 = \frac{K}{M_1},$$

$$A_2 = \frac{K}{M_2},$$

$$B(\boldsymbol{q}) = -\frac{K(\boldsymbol{q})}{\sqrt{(M_1 M_2)}}.$$

Hence the above determinant can be written in the form

$$(A_1-\omega_T^2)(A_2-\omega_T^2)-B^2(\boldsymbol{q}) = 0 \qquad (1.30)$$

which has two distinct solutions as $\boldsymbol{q} \to 0$,

$$\omega_{T1} \simeq \omega_0[1-\alpha(qa)^2] \qquad (1.31)$$

and

$$\omega_{T2} \simeq cq. \qquad (1.32)$$

Here are introduced the following symbols:

$$\omega_0 = \sqrt{\left(\frac{K}{m}\right)}$$

$$m = \frac{M_1 M_2}{M_1 + M_2},$$

$$\alpha = \frac{M_1 M_2}{6(M_1 + M_2)^2},$$

$$c = a\sqrt{\left(\frac{K}{3(M_1 + M_2)}\right)}.$$

The two solutions correspond to the optical and acoustic phonon frequencies, respectively.

Directions of the polarization vectors depend on the plane where the atoms experience vibrational motions. It is easy to see that the plane of vibration (111) is represented by the longitudinal components

$$\boldsymbol{e}_1 \| \boldsymbol{e}_2 \| \quad [111],$$

the transverse components being normal to this direction. A similar definition holds for the remaining planes of vibration.

The elastic force constant, K, can be estimated within the present model as

$$K \cong \frac{48e^{*2}}{v} \tag{1.33a}$$

with e^* and v designating an effective ion charge and the volume of a unit cell, a^3, respectively. The effective charge varies from 0.70 to 0.93 with respect to the actual electron charge. A close agreement is reached between the present model optical frequencies and those observed by infrared transmission technique (see Table 1.5). The present optical mode is represented by $F_1(x, y, z)$ of the space point group O_h which is given by a direct product

$$O_h(m3m) = O(\bar{4}32) \times \mathcal{G}_i(\bar{1}).$$

TABLE 1.5.

Infrared Activity in NaCl *Type Crystals and Elastic Force Constants*

Crystal	M_1+M_2†	e^*/e‡	$\hbar\omega_T$ (cm^{-1})‡	K (dyne/cm)
LiF	7+19	0.87	272	1.96×10^4
NaF	23+19	0.93	190	1.58
NaCl	23+35.5	0.74	164	2.50
NaBr	23+80	0.70	144	2.88
KF	39+19	0.88	172	2.05
KCl	39+35.5	0.81	142	2.50
KBr	39+80	0.76	119	2.77
RbCl	85+35.5	0.84	122	2.77
RbBr	85+80	0.84	95	2.91
AgCl	108+35.5	0.71	118	3.28
AgBr	108+80	0.70	90	3.17

† In proton-mass units. ‡ After Burstein.[11]

Therefore the same selection rule applies for both space point groups. The actual space point group corresponding to a sodium-chloride structure and all related substances is just $O_h(m3m)$.

Using the numerical data from Table 1.5 we can evaluate an average elastic force constant by making equal the present model optical frequency at $\boldsymbol{q} = 0$ with the corresponding frequency as observed in the infrared spectrum. The result is

$$K_{\text{av}} = \frac{1}{11}\sum_{j=1}^{11} K_j = \left(2.58\pm\begin{matrix}0.79\\1.00\end{matrix}\right)\times 10^4 \text{ dyne/cm}. \qquad (1.33\text{b})$$

In general an infrared transmission activity will be very strong for long-wavelength transverse optical phonons. On the other hand, there will be a strong infrared reflection activity in the region separating the longitudinal from the transverse optical phonons.

The longitudinal optical phonon frequency, however, can also be calculated exactly within the present theory on a somewhat phenomenological basis. Indeed, owing to the local forces acting between the positive and negative charges in an ionic crystal (which have been so far neglected within the previous calculation) there will arise an additional force on the vibrating mass. (An elementary treatment of this effect is given by Kittel.[12])

If the relative displacement of the positive and negative ion lattices is denoted by

$$u = u_{(+)} - u_{(-)} \tag{1.34}$$

then the equation of motion for the transverse vibrations is

$$m\ddot{u} + m\omega_T^2 u = 0 \tag{1.35}$$

where the meaning of the used symbols is clear. Now the restoring force on a given ion is no longer only $m\omega_T^2 u$ but actually is given by

$$-m\omega_L^2 u = -m\omega_T^2 u + e^* E_i \tag{1.36}$$

where E_i designates the internal electric field which comes from an overall lattice deformation. This field satisfies the equation

$$D = E_i + 4\pi P = 0 \tag{1.37}$$

with P being the polarization on the ions. Having assumed

$$P = ne^* u \tag{1.38}$$

we arrive at the relation

$$\omega_L^2 = \omega_T^2 + 4\pi \frac{ne^{*2}}{m}. \tag{1.39}$$

Here n denotes the total number of cells per unit volume.

It is worth noting that the two terms appearing on the right side of equation (1.39) are approximately equal to each other owing to the equality

$$\frac{48e^{*2}}{a^3} \cong 4\pi ne^{*2}$$

with n being roughly equal to $4a^{-3}$. Therefore one would expect the following approximate relation to hold:[11]

$$\omega_L \cong \sqrt{2} \cdot \omega_T. \tag{1.40}$$

Here in the present study we shall pay a special attention to the lattice dynamics of potassium dihydrogen phosphate in order to formulate the normal hydrogen vibration modes. It is a striking feature that a close analogy between the ionic crystals and all hydrogen-bonded ferroelectrics exists as far as lattice dynamics is concerned at least. Assuming that the charges are well separated in the form of

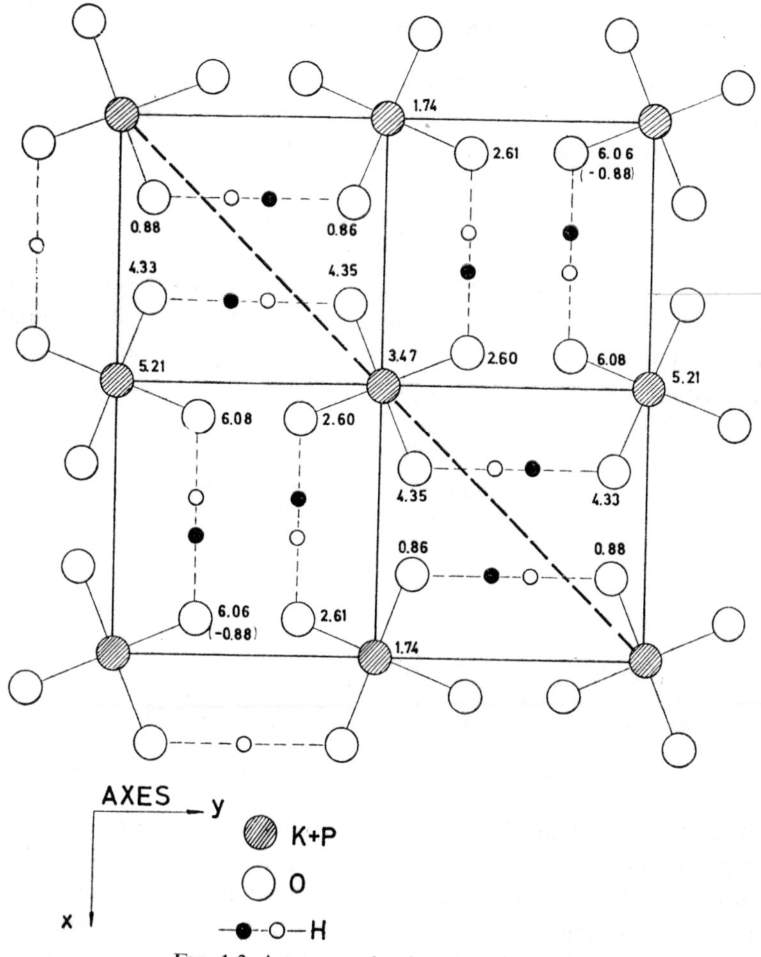

FIG. 1.3. A tetragonal unit cell in KH_2PO_4.

potassium ions being positive on one sublattice and phosphate clusters being negative on the other sublattice we expect to observe the long-wavelength transverse and longitudinal phonons to be given by equations (1.31) and (1.40), respectively, with the same elastic force constant K on the average. The reduced mass in this case is readily calculated as

$$m = \frac{M(K)\cdot M(H_2PO_4)}{M(K)+M(H_2PO_4)} = 27.81 \qquad (1.41)$$

in proton-mass units. Furthermore, if we count four potassium atoms per unit cells as on Fig. 1.3 (and also four phosphate clusters) we can calculate the average number of cells per unit volume,

$$n = \frac{4}{a^2c} = 1.04 \times 10^{22}\ \text{cm}^{-3}$$

where $a = 7.459$, $c = 6.959$ in units 10^{-8} cm. These data are used to estimate an energy corresponding to transverse and longitudinal optical phonons (see Table 1.6). Along with the present analysis also quoted are the experimental figures as reported by Hadži[13] and by

TABLE 1.6.

Energy Corresponding to the Transverse and Longitudinal Optical Phonons of KH_2PO_4

	Hadži[13]	Barker and Tinkham[14]	Present work
$\hbar\omega_T(\text{cm}^{-1})$	123	125 [100]	$123\pm{}^{16}_{27}$
$\hbar\omega_L(\text{cm}^{-1})$	?	175 [001]	$174\pm{}^{23}_{38}$

Barker and Tinkham[14] both using the infrared reflection. As stated explicitly in the latter reference the two phonons are observed in two distinct and mutually orthogonal directions; the transverse optical phonon is detected in both the [100] and [010] directions, whereas the longitudinal optical phonon is detected just in the [001] direction.

This agrees fairly well with the present model. The result of Hadži, however, although being in numerical agreement with that of Barker and Tinkham, is lacking a correct theoretical interpretation; for if the transverse optical phonon were indeed coming from the proton tunneling motion (as assumed by Hadži) then this phonon would be shifted away from the corresponding infrared reflection pattern in a single potassium dideuterium phosphate crystal. However, as observed and reported by Barker and Tinkham both transverse and longitudinal optical phonons are detected from both crystals exactly at the same frequencies and also in the correctly fixed directions. This seems to fit nicely a general theoretical framework towards the observed infrared spectrum.

1.7. Diffuse modes

Any reasonable study of hydrogen-bonded ferroelectrics has to start from an exact space structure, for even a simplified dynamical picture will heavily depend on the assumed space point group. Furthermore, not only the group elements but also the atomic clusters are extremely valuable in evaluating the normal lattice vibration modes.

The crystal structure of potassium dihydrogen phosphate is described in crystallography by a tetragonal symmetry group through a conventional unit cell, the dimensions of the unit cell being $a = b = 7.459$ Å and $c = 6.959$ Å. The (001) plane of this cell is illustrated in Fig. 1.3. Clearly an actual structure consists of two interpenetrating body-centered lattices of the phosphate clusters and two interpretating body-centered potassium lattices, the phosphate and potassium lattices being separated by $c/2$ in the [001] direction. Each phosphate cluster is connected by four-hydrogen bonds to the phosphate clusters on the neighbor sublattice. The hydrogen bonds lie almost parallel to both the [100] and [010] directions. Therefore, one can construct a primitive cell, that is the unit cell having a minimum possible volume, on each sublattice, so that every primitive cell contains four nonequivalent hydrogen atoms. Example of this cell with a minimum volume is illustrated in Fig. 1.4. Here the edges of a new cell are determined by

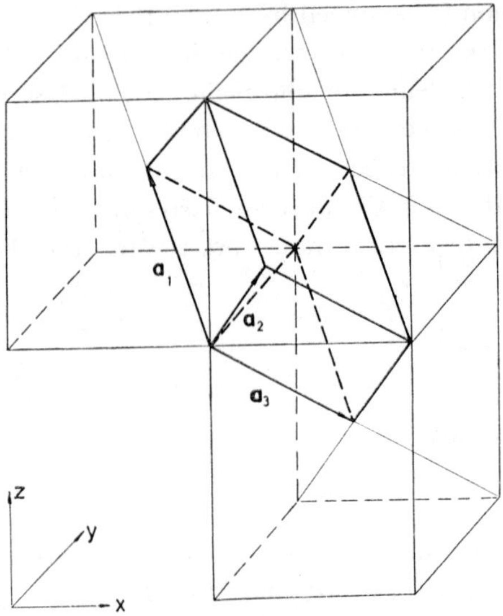

FIG. 1.4. A primitive cell in KH_2PO_4.

the edges of the previous one by

$$\boldsymbol{a}_1 = \left(-\frac{a}{2}, \frac{a}{2}, \frac{c}{2}\right),$$

$$\boldsymbol{a}_2 = \left(\frac{a}{2}, -\frac{a}{2}, \frac{c}{2}\right),$$

$$\boldsymbol{a}_3 = \left(\frac{a}{2}, \frac{a}{2}, -\frac{c}{2}\right). \tag{1.42}$$

So the volume of the new primitive cell is one-half the volume of the previous unit cell,

$$v_{\text{new}} = [(\boldsymbol{a}_1 \times \boldsymbol{a}_2) \cdot \boldsymbol{a}_3] = \frac{a^2 c}{2} = \tfrac{1}{2}\, v_{\text{unit}}.$$

Unfortunately the primitive cell just constructed is inconvenient for the use in any kind of numerical analysis. We will therefore accept the conventionally introduced tetragonal unit cells with the edges being parallel to the crystallographic axes $\boldsymbol{a}$, $\boldsymbol{b}$ and $\boldsymbol{c}$.

One obvious advantage with the selected unit cells is a neat, rational, and illuminative way of presenting the planes and directions. For instance, by following the arguments expressed in Section 1.4 any reciprocal lattice vector $\boldsymbol{G}$, if written in the form

$$\boldsymbol{G} = 2\pi \left(\frac{h}{a}, \frac{k}{a}, \frac{l}{c} \right),$$

is parallel to the direction $[hkl]$ and normal to the plane (hkl). For this reason the directions [100] and [010] are physically equivalent and so all the basic dynamical information can be gathered by performing the experiment merely in the $(h0l)$ plane.

Let us devote our attention in the following exposition to the full study of the relative motions of the phosphate clusters as they oscillate one towards another. Obviously any tetragonal unit cell contains on the average four such clusters; their center-of-mass coordinates with respect to the central phosphate cluster, u, v, and w, are given in Table 1.7. Even a superficial observation of the constellation of heavy

TABLE 1.7.

Center-of-mass Coordinates for Non-equivalent Phosphate Clusters being Placed on the Left, Below, on the Right, and Above Relative to the Central Cluster as Viewed on Fig. 1.3. Components are Given in Direct-lattice Units

	u	v	w
P (left)	0	$-\frac{1}{2}$	$\frac{1}{4}$
P (below)	$\frac{1}{2}$	0	$-\frac{1}{4}$
P (right)	0	$\frac{1}{2}$	$\frac{1}{4}$
P (above)	$-\frac{1}{2}$	0	$-\frac{1}{4}$

masses will suggest that the heavy masses will predominantly experience only low-frequency vibrations at sufficiently low temperatures. As we know, any scattering of thermal neutrons is an ideal means of observing the low-energy lattice modes.

For this reason a coherent nearly-elastic neutron scattering will be analyzed in some details. It is a common fact that the Laue and Bragg equations entirely determine all possible reflections from a given crystal plane, (hkl), although they will not allow to conclude the relative intensities for particular reflections. The latter equations depend on the type of the unit cell as well as on the atomic arrangement within each cluster separately. The structure amplitude for a neutron as scattered by the reciprocal lattice vector $\boldsymbol{G}$ is written

$$F(\boldsymbol{G}) = \sum_{j=1}^{4} f_j \exp\left[i2\pi(hu_j + kv_j + lw_j)\right] \tag{1.43}$$

where f_j designates the atomic factor for a particular phosphate cluster. Since all the atomic factors in the present study happen to be identical then the structure amplitude is reduced to the calculation of the structure factor, $S(\boldsymbol{G})$, which is defined by

$$F(\boldsymbol{G}) = fS(\boldsymbol{G}),$$

$$S(\boldsymbol{G}) = \sum_{j=1}^{4} \exp\left[i2\pi(hu_j + kv_j + lw_j)\right] \tag{1.44}$$

with f being a single atomic factor.

A coherent nearly-elastic neutron scattering (also named the diffuse scattering) has been measured on KD_2PO_4 by Paul *et al.*[15] at various reciprocal lattice points in the plane (010); hence all the relevant vectors are parallel to the directions $[h0l]$ with h and l taking various integer values. Now equation (1.44) yields the most general form for the structure factor,

$$S(\boldsymbol{G}) = 2\left[\exp(i\tfrac{1}{2}\pi) + \cos h\pi \exp(-i\tfrac{1}{2}\pi)\right]. \tag{1.45}$$

The numerical result within the present model can be anticipated in rather general terms in order to establish a set of "selection" rules for the occurrence of a particular diffuse mode. According to the mentioned rules the modes are classified as "weak" or "strong"; these expressions refer to the squared relative structure amplitude with an order of magnitude equal to zero or one, respectively. The observed diffusion modes together with the result of equation (1.45) are presented in Table 1.8. An overall reasonable but not ideal agreement exists between the observed and calculated selection rules, the present departures being marked by (−). One can find a simple and rather evident explanation for the obtained results; namely the neutrons being scattered mostly by a coherent and elastic mechanism from the phosphate clusters having neglected the scattering from the potassium atoms do not reflect precisely the structure and mutual arrangement

TABLE 1.8.

The Observed Diffusion Modes at Various Lattice Points using a Coherent Nearly-Elastic Neutron Scattering on KD_2PO_4 *and Calculated Squared Structure Factors. The Components of* **G** *are Given in Reciprocal-Lattice Units*

G	Experimental[15]	Calculated $(1/16)\,|S(G)|^2$
[002]	weak	1 (−)
[004]	strong	1
[200]	weak	0
[202]	weak	0
[204]	weak	0
[301]	strong	1
[303]	strong	1
[402]	weak	1 (−)

of the clusters as suggested by tetragonal unit cells. (Let us emphasize that the contribution coming from deuterium bonds, although important to some extent for a total coherent effect, also has been neglected in the present model.) Nevertheless, the suggested model, however restricted and simple-minded it appears, will prove a useful framework in relation to the lattice dynamics in order to study the low-frequency motions as experienced by massive phosphate clusters.

For this reason we shall start from the assumption that any motion of hydrogen atoms is strongly coupled to the motion of massive phosphate clusters so that the vibration modes for the two apparently distinguished dynamical systems resemble each other. In particular, should the phosphate clusters experience the low-frequency vibrations accordingly with certain symmetry elements of the space point group $\mathcal{D}_{2d}$, also the hydrogen atoms must experience the same vibrational spectrum with precisely the same vibrational frequencies and polarization vectors. Therefore the motion of the selected phosphate clusters within the (001) plane is expected to be identical with the motion of hydrogen atoms, the latter being strongly coupled to the selected clusters.

1.8. The normal hydrogen vibration modes, I

The Debye temperature, θ, for crystal lattices like those of ferromagnets, antiferromagnets, or hydrogen-bonded ferroelectrics is a crucial physical variable as it determines the upper limit of the validity of a harmonic approximation. In the region above the Debye temperature the harmonic expansion for the lattice potential energy is no longer correct as anharmonic forces become predominant. In the case of ferromagnetism and antiferromagnetism this temperature plays a minor importance since the mentioned phenomena are hardly expressed in terms of lattice dynamics. However, this temperature in the case of hydrogen-bonded ferroelectrics deserves a key position as the whole phenomenon is basically a material part of lattice dynamics.

For crystals with tetragonal unit cells the Debye temperature is estimated from the expression

$$\theta = \frac{hc_{66}}{kd} \tag{1.46a}$$

where d is approximately equal to 7 Å, while c_{66} is the velocity of sound along the tensor elastic compliance component C_{66}^E, k designates Boltzmann's constant. By definition

$$c_{66} = \sqrt{(C_{66}^E/\varrho)}. \tag{1.46b}$$

Here ϱ denotes the density of the material. Using $C_{66}^E \cong 6\times 10^{10}$ dyne/cm^2 and $\varrho \cong 2.338$ g/cm^3 one obtains

$$\theta \cong 110 \text{ K}. \tag{1.46c}$$

Hence the Debye temperature is approximately equal to the transition temperature in KH_2PO_4. Any low-temperature expansion is therefore expected to work in the whole ferroelectric region.

Following arguments of Section 1.5 the lattice hamiltonian can be written in terms of displacement operators referring to the phosphate clusters with respect to the fixed equilibrium positions as observed by neutron scattering. Each displacement operator is represented by a two-valued variable as follows:

$$u_{f\alpha} = b_{f\alpha}^{\dagger} + b_{f\alpha} \tag{1.47}$$

with f and α designating the unit cell and the cluster site within the cell, respectively. The lattice hamiltonian in a transformed form can be written

$$\mathcal{H} = \sum_{\boldsymbol{q}} \left\{ \sum_{\alpha=1}^{n} U b_{\boldsymbol{q}\alpha}^{\dagger} b_{\boldsymbol{q}\alpha} - \sum_{\alpha,\beta=1}^{n} V_{\alpha\beta}(\boldsymbol{q})\, b_{\boldsymbol{q}\alpha}^{\dagger} b_{\boldsymbol{q}\beta} \right.$$
$$\left. -\tfrac{1}{2} \sum_{\beta,\alpha=1}^{n} V_{\alpha\beta}(\boldsymbol{q}) (b_{-\boldsymbol{q}\alpha}^{\dagger} b_{\boldsymbol{q}\beta}^{\dagger} + b_{\boldsymbol{q}\alpha} b_{-\boldsymbol{q}\beta}) \right\} \tag{1.48}$$

where

$$U = \sum_{g,\beta} J(fg;\alpha\beta); \qquad V_{\alpha\beta}(\boldsymbol{q}) \equiv \tfrac{1}{2} J_{\alpha\beta}(\boldsymbol{q});$$

$$J_{\alpha\beta}(\boldsymbol{q}) = J_{\beta\alpha}(\boldsymbol{q}) = \sum_{|\boldsymbol{R}_f - \boldsymbol{R}_g|} J(fg;\alpha\beta) \exp[i\boldsymbol{q}(\boldsymbol{R}_f - \boldsymbol{R}_g)].$$

In the above expression U denotes a total kinetic energy for a given phosphate cluster, whereas $J(fg;\alpha\beta)$ are certain coupling constants.

In general n refers to a total number of protons per unit cells. The sum indicated by $\boldsymbol{q}$ is extended over the first Brillouin zone.

Only two different coupling constants appear in the above hamiltonian, one connecting the phosphate clusters along the equivalent directions [100] and [010] on a given sublattice, the other connecting the clusters along different directions on neighbor sublattices; in particular

$$J[100] = J[010] \equiv K \tag{1.49a}$$

$$J[22\bar{1}] = J[2\bar{2}1] = J[\bar{2}\bar{2}\bar{1}] = J[\bar{2}\bar{2}1] \equiv J. \tag{1.49b}$$

It is obvious that the present choice of the coupling constants is consistent with the case $n = 4$. Moreover, the conservation of energy requires that the introduced quantities should satisfy the condition

$$U = 2J + K. \tag{1.50}$$

The hamiltonian (1.48) is a mathematical bilinear form composed of the boson operators $\boldsymbol{b}^{\dagger}_{q\alpha}$ and $\boldsymbol{b}_{q\alpha}$. To the best of our knowledge it is possible to diagonalize it by an appropriate unitary transformation of the Bogoliubov type.[16] Hence by introducing a new set of boson operators

$$A^{\dagger}_{q\alpha} = \sum_{\beta=1}^{4} u_{\alpha\beta}(\boldsymbol{q})\, b^{\dagger}_{q\beta} - \sum_{\beta=1}^{4} v_{\alpha\beta}(\boldsymbol{q})\, b_{-q\beta},$$

$$A_{q\alpha} = \sum_{\beta=1}^{4} u_{\alpha\beta}(\boldsymbol{q})\, b_{q\beta} - \sum_{\beta=1}^{4} v_{\alpha\beta}(\boldsymbol{q})\, b^{\dagger}_{-q\beta} \tag{1.51}$$

with $u_{\alpha\beta}(\boldsymbol{q})$ and $v_{\alpha\beta}(\boldsymbol{q})$ being some even and real functions of $\boldsymbol{q}$, the original hamiltonian is expressed in the form

$$\mathcal{H} = \sum_{q} \sum_{\alpha=1}^{4} E(\boldsymbol{q};\alpha)\, A^{\dagger}_{q\alpha} A_{q\alpha}. \tag{1.52}$$

Now on applying the equations of motion

$$[A^{\dagger}_{q\alpha}, \mathcal{H}] = -E(\boldsymbol{q};\alpha)\, A^{\dagger}_{q\alpha},$$

$$[A_{q\alpha}, \mathcal{H}] = E(\boldsymbol{q};\alpha)\, A_{q\alpha} \tag{1.53}$$

with

$$[A_{q\alpha}, A^{\dagger}_{q'\alpha'}] = \Delta(q-q')\,\Delta(\alpha-\alpha'),$$

$$\sum_{\beta=1}^{4} [u_{\alpha\beta}(q)\, u_{\alpha'\beta}(q) - v_{\alpha'\beta}(q)\, v_{\alpha\beta}(q)] = \Delta(\alpha-\alpha') \tag{1.54}$$

one obtains

$$\sum_{\beta=1}^{4} \{U - E\,\Delta(\alpha'-\beta) - V_{\alpha'\beta}(q)\}\, u_{\alpha\beta}(q) - \sum_{\beta=1}^{4} V_{\alpha'\beta}(q)\, v_{\alpha\beta}(q) = 0 \tag{1.55}$$

and

$$-\sum_{\beta=1}^{4} V_{\beta\alpha'}(q)\, u_{\alpha\beta}(q) + \sum_{\beta=1}^{4} \{U + E\,\Delta(\beta-\alpha') - V_{\beta\alpha'}(q)\}\, v_{\alpha\beta}(q) = 0; \qquad E \equiv E(q;\alpha). \tag{1.56}$$

The above equations must hold for all values α, $\alpha' = 1, 2, 3$, and 4. Therefore the unique solution for the auxiliary variables $u_{\alpha\beta}(q)$ and $v_{\alpha\beta}(q)$ will exist only if

$$\det \begin{vmatrix} U-E-V(q) & -V(q) \\ -V(q) & U+E-V(q) \end{vmatrix} = 0 \tag{1.57a}$$

where $V(q)$ is a four-by-four matrix with the elements $V_{\alpha\beta}(q)$. Also $U-E$ and $U+E$ are both diagonal matrices of the same dimension. The above determinant has the following explicit form:

$$(E^2-U^2)^4 + \sum_{\beta=2}^{4} P_\beta(q)(E^2-U^2)^{4-\beta} = 0 \tag{1.57b}$$

where the coefficients $P_\beta(q)$ are given by

$$\begin{aligned} P_2(q) &= -8U^2[V_{12}^2(q) + V_{13}^2(q) + V_{14}^2(q)] \\ P_3(q) &= 64U^3 V_{12}(q)\, V_{13}(q)\, V_{14}(q) \\ P_4(q) &= 16U^4 \det |V(q)|. \end{aligned} \tag{1.57c}$$

Each solution to the above determinant will be named the energy of the normal hydrogen vibration mode. For each particular lattice site or phosphate cluster it is given by a general expression where the site suffix varies from one to four. For the first lattice site one obtains

$$\begin{aligned} E_1^2(\boldsymbol{q};1) &= U^2-2U[V_{12}(\boldsymbol{q})+V_{13}(\boldsymbol{q})+V_{14}(\boldsymbol{q})],\\ E_2^2(\boldsymbol{q};1) &= U^2-2U[V_{12}(\boldsymbol{q})-V_{13}(\boldsymbol{q})-V_{14}(\boldsymbol{q})],\\ E_3^2(\boldsymbol{q};1) &= U^2+2U[V_{12}(\boldsymbol{q})+V_{13}(\boldsymbol{q})-V_{14}(\boldsymbol{q})],\\ E_4^2(\boldsymbol{q};1) &= U^2+2U[V_{12}(\boldsymbol{q})-V_{13}(\boldsymbol{q})+V_{14}(\boldsymbol{q})]. \end{aligned} \quad (1.58)$$

Of course, the vibrational energies are site-independent quantities as the reciprocal lattice vector must be parallel to the [001] direction. Hence by having deleted the site suffix the vibrational energies can be written

$$\begin{aligned} E_1^2(\boldsymbol{q}) &= 2JU[1-\exp(iqc/2)],\\ E_2^2(\boldsymbol{q}) &= E_3^2(\boldsymbol{q}) = 2(J+K)\,U,\\ E_4^2(\boldsymbol{q}) &= 2JU[1+\exp(iqc/2)] \end{aligned} \quad (1.59)$$

with $\boldsymbol{q}$ being the unique reciprocal lattice vector that is invariant with respect to all group elements of the space point group $\mathcal{D}_{2d}$,

$$\boldsymbol{q} = (0, 0, q). \quad (1.60)$$

1.9. The normal hydrogen vibration modes, II

The structure and the identification of the normal hydrogen vibration modes from a group-theoretical viewpoint is presented in Table 1.9. Here the pseudo-spin components, $\sigma = (+)$ for each right → left tunneling and $\sigma = (-)$ for each left → right tunneling, are assigned to the four phosphate clusters together with the four corresponding hydrogen bondings. (The labeling of the phosphate clusters agrees with that in Table 1.8.) Of course, the lowest-energy vibration mode, $E_1(\boldsymbol{q})$, is identified by an arbitrary chosen pseudo-spin assignment $\sigma = (+)$ for every one phosphate cluster which corresponds to the representation $B_2(xy; z)$ of the space point group $\mathcal{D}_{2d}$. According to

Kaminow[17] this vibrational energy has to accompany the beta-tunneling mode. The complete pictorial identification of the obtained modes is presented in Fig. 1.5. Here we can observe immediately the

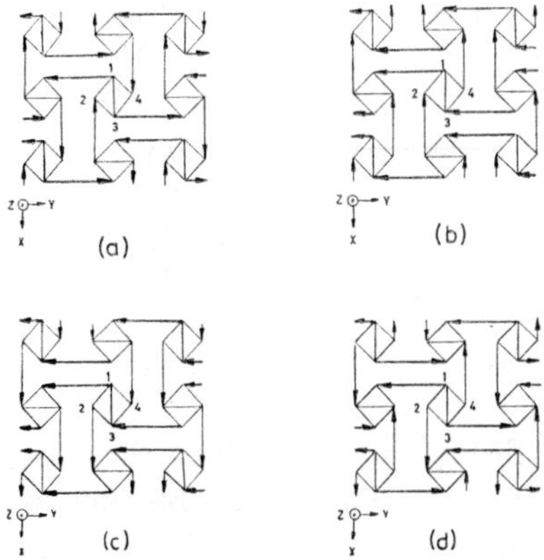

FIG. 1.5. A pictorial representation showing the normal hydrogen vibration modes: (a) $E_1(\boldsymbol{q})$ being represented by $B_2(xy; z)$; (b) and (c) E_2 and E_3 being represented by $E(xz, yz; x, y)$; (d) $E_4(\boldsymbol{q})$ being represented by $A_1(x^2+y^2, z^2)$.

meaning of the lowest-energy vibration mode by applying the operations S_4 and C_2 of the space point group $\mathcal{D}_{2d}$. Indeed, these two operations are identical with the rotation through an angle $\pi/2$ with a mirror reflection on the (001) plane and the rotation through an angle π, respectively. By an eye-circling vision so as to have the central phosphate cluster to the left one observes an exactly invariant pattern of this vibration mode, the pattern being represented by the group characters -1 and 1, respectively, as indicated in the last two columns in Table 1.9. It is a Raman active mode.

A similar interpretation holds for the next two vibration modes whose energy is a $\boldsymbol{q}$ independent quantity and is given by the same expression, $E_2 = E_3$. Both modes are represented by a two-dimensional representation $E(xz, yz; x, y)$ of the space point group $\mathcal{D}_{2d}$. In the quoted paper by Kaminow this vibrational energy has to accompany the gamma-tunneling modes. Here again we have the opportunity of presenting a pictorial view of the two essentially identical modes, as

TABLE 1.9.

The Normal Hydrogen Vibration Modes of a Single KH_2PO_4 Crystal

Cluster pseudo-spin assignment, σ	Vibrational energy	Representation of the space point group $\mathcal{D}_{2d}$		Group character	
P1 P2 P3 P4		Present work	References 17–18	S_4	C_2
+ + + +	$E_1(\boldsymbol{q})$	$B_2(xy; z)$	Γ_4	−1	1
+ + − −	$E_2(\boldsymbol{q})$	$E(xz; yz; x, y)$	Γ_5	0	−2
+ − − +	$E_3(\boldsymbol{q})$	$E(xz; yz; x, y)$	Γ_5	0	−2
+ − + −	$E_4(\boldsymbol{q})$	$A_1\,(x^2+y^2, z^2)$	Γ_2	1	1

seen in Fig. 1.5. In distinction to the others the present modes are both infrared and Raman active. This fact enables one to estimate the magnitude of the characteristic energy which is expected to agree with the energy of the transverse optical phonon as $\boldsymbol{q} \to 0$. Therefore

$$E_2 = E_3 = \sqrt{\{2(J+K)\,U\}} = \hbar\omega_T(0). \tag{1.61}$$

The present modes have the pattern as represented by the group characters 0 and −2 for the operations S_4 and C_2, respectively; see again Table 1.9.

The highest-energy vibration mode, $E_4(\boldsymbol{q})$, is represented by the one-dimensional representation $A_1(x^2+y^2, z^2)$ of the space point group $\mathcal{D}_{2d}$. According to the cited work by Kaminow this vibrational energy has to accompany the alpha-tunneling mode, the pattern being represented by the group character 1 for both operations, S_4 and C_2, as indicated in the last two columns in Table 1.9. It is a Raman active mode.

Experimental observation of the Raman activity for the present vibration modes in KH_2PO_4 together with the theoretical anticipation are collected in Table 1.10. Here one may notice an excellent agreement between the two sets of presented data. Of course, the present theory, as well as that of Shur,[(22)] is based on the group concepts; so strictly speaking it is expected to be correct only at absolute zero where the

TABLE 1.10.

Observed and Calculated Vibrational Energies, cm^{-1}, *of a Single* KH_2PO_4 *Crystal*

Vibration modes, $q = 0$	Observed Raman spectrum			Calculated energy	
	Reference 19	Reference 20	Reference 21	Shur's method[22]	Present method
$B_2(xy; z)$	85	—	80	5	0
$E(xz, yz; x, y)$	80; 92; 112; 182	96; 114; 190	95; 113; 190	68; 99; 118	123^{+16}_{-27}
$A_1(x^2+y^2, z^2)$	357	363	360	—	$331^{+17‡}_{-14}$

‡ Using a self-consistent molecular field approximation, Chapter 2: $U = \frac{1}{2}H_x = 187$; $J = 147^{+15}_{-12}$; $K = 107^{+22}_{-30}$ in units cm^{-1}; H_x designates the high-temperature transverse field.

assumed symmetry elements as being imposed by the real crystal structure are fully realized. For this reason any high-temperature observation, however it might seem to agree with the theoretical anticipation, still has to be confirmed by a low-temperature observation. Of course, one could bring the vibration modes to a high-temperature region by taking into account the deleted terms in the original hamiltonian. However, this would be an extraordinary uncertainty as to a growing influence of both the dynamical and kinematical interactions with the increasing temperature, these interactions being an essential aspect of the pseudo-spin method. In addition one has to bear in mind a serious limitation of the harmonic approximation if applied to any temperature region which exceeds the Debye temperature.

It is clear now that only the lowest-energy vibration mode, $E_1(q)$, associated with the representation $B_2(xy; z)$ of the space point group $\mathcal{D}_{2d}$ has explicitly the probability of behaving in the way which is required for a soft ferroelectric mode.

References

1. C. KITTEL, *Introduction to Solid State Physics*, chap. I, John Wiley & Sons, Inc., New York, 1967.
2. R. A. LEVY, *Principles of Solid State Physics*, chap. 2, Academic Press, New York and London, 1968.
3. L. D. LANDAU and E. M. LIFSHITZ, *Statistical Physics*, chap. XIII, Pergamon Press, London, 1959.
4. G. YA. LYUBARSKII, *The Application of Group Theory in Physics*, chaps. I–V, Pergamon Press, London, 1960.
5. E. B. WILSON, J. C. DECIUS, and P. C. CROSS, *Molecular Vibrations*, Appendix X, McGraw-Hill Book Co., Inc., New York, 1955.
6. M. TINKHAM, *Group Theory and Quantum Mechanics*, Appendix B, McGraw-Hill Book Co., Inc., New York, 1964.
7. V. I. SMIRNOV, *Higher Mathematics*, Vol. III, Part One, sections 52, 53, and 76–79, Nauka, Moscow, 1969.
8. S. MITRA, *Solid State Physics*, Vol. 13, ed. by F. SEITZ and D. TURNBULL, Academic Press, New York, 1962.
9. F. LOUDON, *Adv. Phys. (Phil. Mag. Suppl.)* **13,** 423 (1964).
10. J. M. ZIMAN, *Electrons and Phonons*, chap. I, Oxford University Press, London, 1963.
11. E. BURSTEIN, Infrared and Raman Lattice Vibration Spectra. In: *Lattice Dynamics*, ed. by R. F. WALLIS, Pergamon Press, London, 1965. Also in: *Dynamical Processes in Solid State Optics*, ed. by R. KUBO and H. KAMIMURA, W. A. BENJAMIN, Inc., New York, 1967.
12. C. KITTEL, *Quantum Theory of Solids*, chap. 3, John Wiley & Sons, Inc., New York, 1967.
13. D. HADŽI, *J. Chem. Phys.* **34,** 1445 (1961).
14. A. S. BARKER and M. TINKHAM, *J. Chem. Phys.* **38,** 2257 (1963).
15. G. L. PAUL, W. COCHRAN, W. J. L. BUYERS, and R. A. COWLEY, *Phys. Rev.* **2,** B 4603 (1970).
16. N. N. BOGOLIUBOV, *J. Phys. USSR* **9,** 23 (1947); N. N. BOGOLIUBOV, in: *Izbranie Trudi*, Vol. II, p. 242, ed. by Naukova Dumka, Kiev, 1970. [English trans.: *Lectures on Quantum Statistics*, Vol. I, MacDonald Technical and Scientific, London, 1968.]
17. I. P. KAMINOW, *Phys. Rev.* **138,** A 1539 (1965).
18. H. MONTGOMERY and G. L. PAUL, *Proc. Roy. Soc.* (Edinburgh) **70,** 107 (1971).
19. B. LAVRENČIĆ, I. LEVSTEK, B. ŽEKŠ, R. BLINC, and D. HADŽI, *Chem. Phys. Lett.* **5,** 441 (1970).
20. C. Y. SHE, T. W. BROBERG, and D. F. EDWARDS, *Phys. Rev.* **4,** B 1580 (1971).
21. D. K. AGRAWAL and C. H. PERRY, in: *Light Scattering in Solids*, ed. by M. BALKANSKI, Flammarion, Paris, 1971.
22. M. SHUR, *Fiz. Tverd. Tela* **8,** 57 (1966). [English trans.: *Soviet Phys. — Solid State* **8,** 43 (1966).]

CHAPTER 2

Phase Transitions

2.1. Magnetism, I

The study of magnetism will be greatly simplified if this phenomenon is sorted as paramagnetism, ferromagnetism, antiferromagnetism, and ferrimagnetism depending on the mutual order and arrangement of magnetic ions.

1. PARAMAGNETISM is the property of having a positive magnetic susceptibility, χ, which occurs in numerous anorganic materials, transition metals and organic compounds. By definition

$$\chi = \frac{M}{H} \tag{2.1}$$

with M being the total magnetic moment per unit volume (cm^3) and H being the magnetic field strength expressed in oersteds. A classical statistical theory of an ensemble of dipole moments, μ, in thermal equilibrium in a magnetic field is done by Langevin. The result at a given temperature, T, is[1]

$$M = N\mu L(x),$$
$$x = \frac{\mu H}{kT}, \tag{2.2}$$

where N is the total number of atoms per unit volume. The Langevin function appearing in the above expression is defined by

$$L(x) = \coth x - \frac{1}{x} = \frac{\exp x + \exp(-x)}{\exp x - \exp(-x)} - \frac{1}{x}. \tag{2.3}$$

At high temperatures and in weak magnetic fields the following condition applies:

$$\mu H \ll kT, \quad \text{or} \quad x \ll 1,$$

so the paramagnetic susceptibility reduces to the Curie law

$$\chi = \frac{N\mu^2}{3kT} = \frac{C}{T} \tag{2.4}$$

with C being the Curie constant. A quantum-statistical theory for atoms with the total angular momentum spin quantum number S yields a slightly different result,

$$M = N\mu S B_S\left(\frac{\mu S H}{kT}\right),$$

$$\mu = g\mu_B, \tag{2.5}$$

where g is the Landé factor, μ_B is the Bohr magneton. By definition the present Brillouin function (see Fig. 2.1) is

$$B_S(x) = \frac{2S+1}{2S}\coth\frac{(2S+1)}{2S}x - \frac{1}{2S}\coth\frac{x}{2S}. \tag{2.6}$$

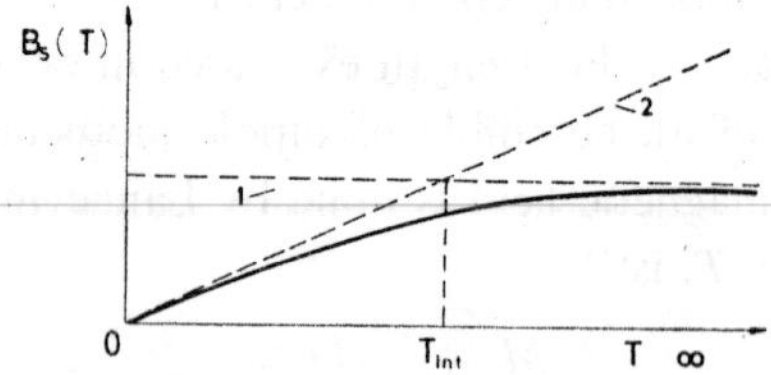

FIG. 2.1. Schematic representation of the Brillouin function: (1) high-temperature limit $B_S(T \to \infty) = 1$; (2) low-temperature limit $B_S(T \to 0) = (S+1)\,x/3S$, where $x = g\mu_B\, SH/kT$. The intersection point: $T_{\text{int}} = g\mu_B(S+1)\,H/3k$.

At high temperatures and in weak magnetic fields the paramagnetic susceptibility becomes

$$\chi = NS(S+1)\frac{\mu^2}{3kT}\,. \tag{2.7}$$

(It is worth emphasizing that the two theories, one being classical the other quantum-mechanical, meet together whenever S becomes unlimitedly large.) Observed physical quantities jointly with the complete pseudo-spin assignment for a number of solid elements and compounds are collected in Table 2.1.

TABLE 2.1.

Curie Constants, Effective Bohr Magneton Numbers and Pseudo-Spin Magnitude of Paramagnetic Materials
(First five columns after Bozorth[2]; Θ is constant in Curie–Weiss law)

Substance	Range of validity of the Curie–Weiss law, K	C	Θ, K	n_{eff} per formula unit‡	Pseudo-spin S
Co	1500 to 1720	1.24	1400	3.15	1
CoF_2	—	2.90	−44	4.81	2
CoO	> 300	3.23	290	5.1	2
Dy	> 150	14.6	150	10.8	5
EuS	—	6.81	6	7.38	3
Fe	> 1100	1.23	1093	3.14	1
FeF_2	—	3.88	−117	5.57	2
Gd	> 300	7.48	302	7.73	$\frac{7}{2}$
K_2MnO_4	—	0.383	−7	1.75	$\frac{1}{2}$
MnF_2	> 90	~4.10	−92	5.7	$\frac{5}{2}$
Ni	> 950	0.402	538	1.79	$\frac{1}{2}$
NiF_2	> 75	1.34	−97	3.27	1
Tb (85%)	—	10.0	205	~9.0	~4
UO_2	—	1.06	−185	2.92	1
$ZnCo_2O_4$	> 100	0.62	−20	2.2	$\sim\frac{1}{2}$
$ZnCr_2S_4$	> 100	3.34	10	5.17	2

‡ By definition

$$n_{eff} = g\sqrt{\{S(S+1)\}}$$

where $g = 2$ for most chemical elements, see Urşu.[3]

2. FERROMAGNETISM is the property of possessing a characteristic spontaneous magnetization after having applied a magnetic field at temperatures below a certain temperature T_c, named the Curie point. Here the following physical models are developed.

(a) *Weiss molecular field.* P. Weiss (1907) pointed out that ferromagnetism can be understood by assuming that each compound consists of tiny domains where all domains are magnetized. In an apparently nonmagnetical piece of iron the direction of magnetization of individual domains is distributed randomly over the lattice volume so that the total magnetic moment vanishes. The process of magnetizing consists of directing the magnetization of individual domains along an applied field in such a way as to make unchanged the intensity of magnetization of individual domains. Measurements show that the magnetization has a maximum value at absolute zero and then it decreases as temperature is increased up to a certain point. Therefore the transition (Curie) temperature is defined as the point which divides every domain in two phases, one completely ordered (ferromagnetic), the other completely disordered (paramagnetic). Below the transition temperature where the magnetization is different from zero the two phases exist together, whereas above it where the magnetization of the domains vanishes owing to the thermal averaging only the disordered phase appears. The ordered phase alone exists only at the absolute zero.

According to Weiss's domain theory an effective magnetic field acting on the individual dipoles consists of the applied field added to an internal (molecular) field which is proportional to the total magnetization,

$$H_{\text{eff}} = H_{\text{app}} + \lambda M. \tag{2.8}$$

The magnetization is given by the Langevin function as follows:

$$M = N\mu L\left(\frac{H_{\text{app}} + \lambda M}{kT}\right) \tag{2.9}$$

where a nonvanishing solution for M if H_{app} tends to zero exists so long as $T \leqslant T_c$. Here the transition (Curie) temperature is given by

$$T_c = \frac{N\mu^2}{3k}\lambda. \tag{2.10}$$

For a disordered phase, $T > T_c$, the susceptibility becomes

$$\chi = \frac{C}{T-T_c} \tag{2.11a}$$

with the Curie constant

$$C = \frac{N\mu}{3k}. \tag{2.11b}$$

These two equations are named the Curie–Weiss law. Keeping the constant C unchanged the above law is usually written in the form

$$\chi = \frac{C}{T-\Theta} \tag{2.11c}$$

where Θ, named the paramagnetic Curie temperature, is experimentally determined by measuring the molar susceptibility for a large number of substances (see Table 2.1). It is observed, however, that Θ is slightly greater than T_c when measured at temperatures well above T_c.

(b) *Ising model*. Ising[4] replaced the molecular field assumption with an idea that the interaction between a pair of atoms j and k has the form

$$V_{jk} = -2JS_{jz} = S_{kz} \tag{2.12}$$

where S_{jz} and S_{kz} are essentially classical variables and J is a coupling constant linking the nearest neighbors. The spontaneous magnetic moment of a square Ising lattice assuming the variables S_{jz} and S_{kz} are quantum-mechanical spins of magnitude one-half is calculated exactly by Onsager,[5] Yang,[6] and Huang.[7] It is equal to zero above the Curie temperature whereas below it has the value

$$M = N\mu\left\{\frac{\cosh^2(2J/kT)}{\sinh^4(2J/kT)}\left[\sinh^2(2J/kT)-1\right]\right\}^{1/8}. \tag{2.13}$$

The Curie temperature is determined by

$$\sinh \frac{2J}{kT_c} = 1$$

or equivalently

$$\frac{J}{kT_c} = 0.4407. \tag{2.14}$$

The properties of three-dimensional Ising lattices are not known accurately at this time although there are a number of approximate solutions. Some of them are in agreement with experimental evidence, the others are not.

(c) *Heisenberg exchange interaction.* Heisenberg[8] replaced the molecular field assumption with an idea that the interaction between neighbor atoms has the form

$$V = -2J\boldsymbol{S}_j\boldsymbol{S}_k \tag{2.15}$$

where $\boldsymbol{S}_j$ and $\boldsymbol{S}_k$ are quantum-mechanical spin variables and J is the exchange energy, or exchange coupling, or exchange integral. This quantity is a function of the distance with respect to a relative position of the two atoms. The present problem has not been solved exactly except for a low-temperature region. However, the high-temperature region including the critical point can be obtained by a number of approximations. A usual approximation consists of considering the nearest-neighbor interactions for which all states of the crystal with a given lattice spin have the same energy. Having made the following replacements

$$\begin{aligned} \mu &\to g\mu_B S \\ \mu^2 &\to (g\mu_B)^2\, S(S+1) \\ L(x) &\to B_S(x) \\ \lambda &\to \frac{2zJ}{N(g\mu_B)^2} \end{aligned} \tag{2.16}$$

we can take over the Weiss molecular field as follows:

$$M = Ng\mu_B S B_S(x) \tag{2.17a}$$

where

$$x = \frac{g\mu_B}{kT} S(H+\lambda M). \tag{2.17b}$$

Here z denotes the number of nearest neighbors encircling a given ferromagnetic atom.

The Curie temperature and susceptibility for $T > T_c$ are readily obtained using the same approximation,

$$T_c = \frac{2zJ}{3k} S(S+1),$$

$$\chi = \frac{C}{T-T_c},$$

$$C = \frac{4N\mu_B^2}{3k} S(S+1). \tag{2.18}$$

(d) *XY model.* This model consists of replacing the Weiss molecular field with an interaction of a purely quantum-mechanical nature,

$$V_{jk} = -2J(S_{jx}S_{kx}+S_{jy}S_{ky}). \tag{2.19}$$

The present problem is very laborious and there are only approximate solutions obtained for certain cubic lattices assuming nearest-neighbor interactions. Here we quote the result for the transition temperature on the pseudo-spin one-half as reported by Betts *et al.*[9] (see Table 2.2).

TABLE 2.2.

Transition Temperature of an XY *Model*
(after Betts *et al.*[9])

Lattice	Cubic P	Cubic I	Cubic F
kT_c/J	2.02	2.90	4.52

2.2. Magnetism, II

Besides the substances whose lattice structure consists of a single lattice also there are substances whose magnetic ions can be divided into equivalent or nonequivalent sublattices.

3. ANTIFERROMAGNETISM is referred to those materials where magnetic ions are divided into equivalent sublattices which become magnetized in an antiparallel arrangement below a certain temperature T_c. The antiparallel arrangement occurs spontaneously owing to a large value of the exchange energy. If we introduce two simple sublattices, say A and B, with nearest-neighbor interactions then we may write two effective fields as follows:

$$\begin{aligned} H_{\text{eff}}(A) &= H_{\text{app}} - \lambda M(B), \\ H_{\text{eff}}(B) &= H_{\text{app}} - \lambda M(A), \end{aligned} \tag{2.20}$$

with λ being the same constant as in the case of ferromagnetism except that J here is essentially negative. The susceptibility for $T > T_c$ can be written

$$\chi = \frac{C}{T-\Theta} \tag{2.21a}$$

where

$$C = \frac{N(g\mu_B)^2 S(S+1)}{3k}. \tag{2.21b}$$

Here we assume that the total lattice is built up of interpenetrating identical cubic sublattices A and B. The two introduced temperatures are related by

$$\Theta = \pm T_c$$

depending on whether the interactions are ferromagnetic or antiferromagnetic. The experimental data on most antiferromagnetic materials can be fitted nicely to the above equations. However, the value of Θ/T_c instead of being -1, as predicted by a simple model, varies within large limits starting from $+0.7$ as observed for MnAs down all the way to -5.0 as observed for MnO (see Van Vleck[10] and Smart.[11]) A number of substances with the described properties are presented in Table 2.3.

TABLE 2.3.

Antiferromagnetic or Néel Points, Θ, and the Structure Type Studied by Neutron Difraction (after Bozorth[2])

Substance	Θ, K	Structure type	Magnetic unit cell in terms of cryst. cell	Θ, K from diffraction
FeF_2	79	SnO_2 tetragonal	*a*, *c*	90
FeO	190	NaCl f.c. cubic	2*a*	—
CoF_2	38	SnO_2 tetragonal	*a*, *c*	45
CoO	291	NaCl f.c. cubic	2*a*	—
NiF_2	73	SnO_2 tetragonal	*a*, *c*	83
NiO	520	NaCl f.c. cubic	2*a*	—
Cr	475	b.c. cubic	—	$\cong$ 475
Mn	100	cubic	*a*	$\cong$ 100

4. FERRIMAGNETISM is referred to those materials where magnetic ions are divided into nonequivalent sublattices which become magnetized in an antiparallel array below a certain temperature. A ferrimagnetic order is realized in numerous ferrites whose general chemical formula is Fe_2O_3MO with M replacing the ion of a divalent metal (e.g. Mn, Co, Ni, Cu, Pb, Mg, Zn, Cd, or $Fe^{2(+)}$). The structure of ferrites is identical to that of magnetite, Fe_3O_4, except that one of the Fe ions is replaced by an ion of the mentioned divalent metals (see Fig. 2.2).

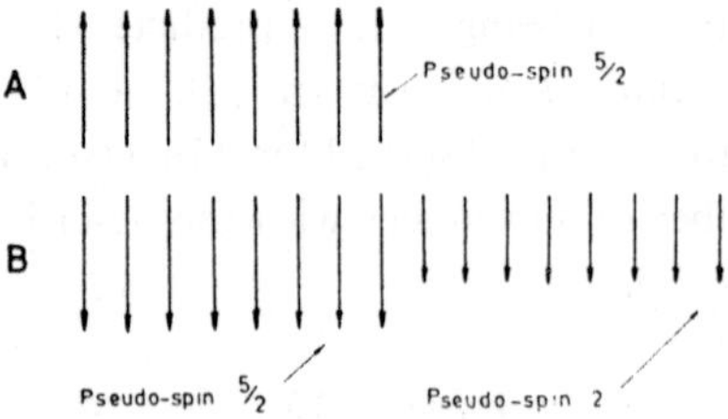

FIG. 2.2. Schematic representation of the spinel structure in magnetite, Fe_3O_4: (A) Tetrahedral sublattice with eight ferromagnetic ions $Fe^{3(+)}$ each having a pseudo-spin 5/2; (B) octahedral sublattice with sixteen ferromagnetic ions: eight having the pseudo-spin 5/2 ($Fe^{3(+)}$) and another eight having the pseudo-spin 2 ($Fe^{2(+)}$).

A basic dynamical theory of the ferrites from which all the contemporary theories on ferrimagnetism originate is established by Néel.(12) It turns out that the magnetic susceptibility is a complicated function of temperature and for most experimental data it can be written

$$\frac{1}{\chi} = \frac{T}{C} + \frac{1}{\chi_0} - \frac{\sigma}{T-\Theta} \tag{2.22}$$

where χ_0, σ, Θ are constants depending on a particular ferrimagnetic compound, and C is the atomic Curie constant of an Fe ion whose

TABLE 2.4.

Ferrites as Represented by Néel Law (after Néel(12))

Substance	$\frac{1}{\chi_0}$	σ	Θ, K	Range of the validity of Néel law, °C
Fe_2O_3MgO I	296.7	14,700	601.8	420–720
Fe_2O_3MgO II	259.3	2486	705.2	480–720
Fe_2O_3PbO a	298.2	10,610	750.0	550–720
Fe_2O_3PbO b	305.8	19,630	708.6	550–750
Fe_2O_3CuO	292.7	9970	744.1	530–710
Fe_2O_3NiO	239.9	10,600	881.0	680–740
Fe_2O_3CdO f	139.8	18,600	555.3	450–750
Fe_2O_3CdO p	158.7	37,100	384.0	450–750
Fe_2O_3ZnO	197.0	300,000	−1210	30–700

value is 4.4. A number of measurements as quoted by Néel are presented in Table 2.4. It is easily seen that a high-temperature limit of the Néel law approaches the usual Curie–Weiss law of antiferromagnetism for most ferrimagnetic materials. Indeed by observing the data on Table 2.4 one can conclude that the introduced parameters χ_0, σ, and Θ are related one to another by an inequality of the kind

$$\frac{\Theta}{\chi} \gg \sigma. \tag{2.23}$$

This enables one to write the susceptibility as

$$\chi \cong \frac{C(T-\Theta)}{(T-T_1)(T-T_2)} \tag{2.24}$$

where $T_{1,2}$ are solutions to the equation

$$T^2 - \left(\Theta - \frac{C}{\chi_0}\right)T - C\frac{\Theta}{\chi_0} = 0.$$

There follows

$$T_1 = \Theta, \qquad T_2 = -\frac{C}{\chi_0},$$

$$\chi = \frac{C}{T+\Theta'}, \tag{2.25}$$

with $\Theta' = C/\chi_0$ being of the order of magnitude of Θ.

2.3. The Ising model with a parallel field

The Ising model does not correctly reflect the dynamical aspects of ferromagnetism which stem from quantum-mechanical commutation relations for different spin components. Therefore we must expect that the Ising model is incorrect at low temperatures where the energy becomes a quadratic function of the components S_x and S_y. Hence we should apply the Heisenberg model in this region in order to enable the system to achieve an equilibrium state in the form of spin waves. On the other hand, there are no spin waves in the Ising model where the thermal equilibrium is destroyed only by the reverse of the individual spins. The present deficiency of the Ising model is immaterial in the vicinity and above the Curie temperature where the statistical counting of states has an outstanding value in comparison to the non-statistical factors of influence.

Let us consider the pseudo-spin one-half with the internal energy of the system including here a constant magnetic field H as

$$\mathcal{H} = -\sum_{\langle jk \rangle} J_{jk} S_{jz} S_{kz} - g\mu_B H \sum_j S_{jz} \tag{2.26}$$

where sums are taken over all interaction pairs and all lattice sites. The symbol $\langle jk \rangle$ is used to indicate that each interaction pair is taken only once into account. Let the total number of relative orientations parallel (up) or antiparallel (down) be N_p and N_a, respectively. Clearly the relative ratio is $n_p = N_p/N$ and $n_a = N_a/N$, respectively. Here $N = N_p + N_a$.

Using a molecular-field approximation the relative ratio becomes a total magnetic moment-dependent quantity as follows:

$$n_p = \tfrac{1}{2}(1+X), \qquad n_a = \tfrac{1}{2}(1-X), \tag{2.27a}$$

with

$$X = \frac{M}{Ng\mu_B}, \qquad n_p + n_a = 1. \tag{2.27b}$$

Here M designates the total magnetization at a given temperature. According to these equations every individual pseudo-spin is replaced by

$$S_{jz} \to \tfrac{1}{2}(n_p - n_a), \tag{2.27c}$$

so the internal energy of the system, per unit volume, can be written

$$-\mathcal{H} = \tfrac{1}{4}\zeta NJ(n_p - n_a)^2 + \frac{N}{2} g\mu_B H(n_p - n_a)$$

where ζ designates the number of nearest neighbors.

According to our earlier assumption the present approximation is valid only in a high-temperature region where the statistical counting of individual spin states becomes essential. It is clear that the internal energy is a total magnetic moment-dependent quantity,

$$-\mathcal{H} = \tfrac{1}{4}\zeta NJX^2+\tfrac{1}{2}HM. \tag{2.28}$$

We shall proceed to formulate the transition temperature as an exchange integral-dependent quantity by calculating the Helmholtz free energy, F, and the entropy of the system, S. By definition

$$F = \mathcal{H}-TS,$$

$$S = k\ln W, \tag{2.29}$$

$$W = (n_p)^{-Nn_p}(n_a)^{-Nn_a}, \tag{2.30}$$

the latter quantity being the probability for the system to have a given macroscopic state. Hence

$$S = -kN(n_p\ln n_p+n_a\ln n_a) = kN[\ln 2-\tfrac{1}{2}(1+X)\ln(1+X) - \tfrac{1}{2}(1-X)\ln(1-X)]. \tag{2.31}$$

By the second law of thermodynamics the free energy of the system has a minimum in its equilibrium state at a given temperature and under given external constraints. Therefore we search for this minimum provided that M varies as a free parameter,

$$\frac{\partial F}{\partial M} = \frac{\partial F}{\partial X} = 0. \tag{2.32}$$

Having performed the indicated differentiation we arrive at the following equation:

$$X = \tan h\left[\frac{g\mu_{\mathrm{B}}}{N2kT}\left(H+\frac{\zeta JN}{g\mu_{\mathrm{B}}}X\right)\right]. \tag{2.33}$$

The above self-consistent equation relates the total magnetic moment to the external magnetic field. The transition temperature, T_c, is obtained by taking the limit where H, M and X tend to zero simultaneously. Hence

$$kT_c = \tfrac{1}{2}\zeta J. \tag{2.34}$$

2.4. Ferroelectricity

Ferroelectricity is the occurrence of a spontaneous electric polarization as an intrinsic property of the crystal. The spontaneous polarization occurs whenever the energy gained by the interaction of electric dipoles induced by a local field is larger than the energy required to produce the dipoles, this quantity being a deformation energy according to a Hook's law approximation. All ferroelectrics are phenomenologically identified by the following properties: spontaneous polarization, temperature dependence, domain structure, isotope effect, symmetry change, and dynamical behavior.

Ferroelectricity occurs only below a certain point named the Curie temperature. (In some materials it occurs only within a narrow temperature interval, the two limiting points being the upper and lower Curie temperatures.) Above the Curie temperature the material behaves like an ordinary dielectric.

There is a certain similarity between ferromagnetism and ferroelectricity in the sense that not the entire volume of a given ferroelectric is polarized in the same direction. The volume is divided into domains, each of them having the polarization which points to one of the basic directions of the lattice.

It is observed in a KH_2PO_4 type of ferroelectrics (also named KDP) that the transition temperature, T_c, is almost doubled upon deuteration. Also the spontaneous polarization is increased when measured at temperatures sufficiently below T_c. This property, usually named the isotope effect, has a profound influence on the dynamical behavior of the light particles (protons, deuterons). The enumerated properties,

TABLE 2.5.

Dielectric Properties of Principal Groups of Ferroelectrics
(after Kanzig[14] and Jona and Shirane[15])

Substance	International name	Principal transition temperature	Spontaneous polarization, $\mu C/cm^2$
$(NH_2CH_2COOH)_3 \times H_2SO_4$	Triglycine sulfate	49 °C	2.8 (20 °C)
KH_2PO_4	Potassium dihydrogen phosphate	122 K	4.75 (96 K)
KD_2PO_4	Potassium dideuterium phosphate	223 K	4.83 (180 K)
$BaTiO_3$	Barium titanate	120 °C	26.0 (23 °C)
$PbTiO_3$	Lead titanate	490 °C	50.0 (23 °C)
$NaNO_2$	Sodium nitrite	164 °C	6.4 (143 °C)
$NaKC_4H_4O_6 \times 4H_2O$	Rochelle salt	+24 °C −18 °C	0.25 (maximum at 5 °C)
$NaKC_4D_4O_6 \times 4D_2O$	Deuterated Rochelle salt	+ 35 °C −22 °C	0.35 (maximum at 6 °C)

according to Megaw,[13] Kanzig,[14] and Jona and Shirane,[15] are presented in Table 2.5 for principal groups of ferroelectrics. It is necessary to emphasize that electric dipoles exist in each material, some of them being induced the others being permanent, (see Table 2.6). A given ferroelectric phase transition is always accompanied by the symmetry change as illustrated in Table 2.7.

There are two kinds of ferroelectrics as to the behavior which they exhibit in a close neighborhood of the transition temperature. One is called the displacive kind; it is identified by the presence of an optical phonon which vanishes as the reciprocal lattice vector tends to zero in the critical region,

$$\boldsymbol{q} \to 0, \qquad T \to T_0.$$

TABLE 2.6.

Dipole Structure, Type of Phase Transition, and Pseudo-spin Magnitude for Principal Groups of Ferroelectrics

Substance	Dipole structure	Type of phase transition	Formula units, n	Total pseudo-spin, S‡
TGS	$(NH_3^+CH_2COO^-)$ $(NH_3^-CH_2COOH)_2$ $\cdot SO_4^{2(-)}$	order–disorder	2	$\frac{3}{2}$
KDP	$K^+H_2PO_4^-$	order–disorder	4	$\frac{15}{2}$
Deuterated KDP	$K^+D_2PO_4^-$	order–disorder	4	$\frac{15}{2}$
$BaTiO_3$	$Ba^{2(+)}TiO_3^{2(-)}$	displacive	1	$\frac{1}{2}$
$PbTiO_3$	$Pb^{4(+)}TiO_3^{4(-)}$	displacive	1	$\frac{1}{2}$
$NaNO_2$	$Na^+No_2^-$	order–disorder (?)	?	?
RS	?	order–disorder	4	$\frac{15}{2}$
Deuterated RS	?	order–disorder	4	$\frac{15}{2}$

‡ By definition $S = (2^n-1)/2$.

This kind is also named the soft-mode phase transition. The other kind is called the order–disorder phase transition; it is identified by the presence of a double-minimum potential field. Examples for the two kinds of ferroelectrics include barium titanate and KDP, respectively. There is a close correspondence between the present definition of the two kinds of ferroelectrics and the conventionally used Ehrenfest's classification, on one hand, and between the present definition and the dielectric nature, on the other hand, see Table 2.8. The first kind of ferroelectrics is also identified by two critical points, T_0 and T_c. The former point refers to the temperature at which the soft-mode optical phonon vanishes for $\boldsymbol{q} = 0$, the latter point refers to the

TABLE 2.7.

Lattice Structure of Principal Groups of Ferroelectrics

Substance	Ordered phase	Disordered phase
Triglycine sulfate References 1, 2, 3	Monoclinic $C_2(2)$	Monoclinic $C_{2h}(2/m)$
KH_2PO_4 References 1, 2, 3	Orthorhombic $C_{2v}(mm2)$	Tetragonal $D_{2d}(\bar{4}2m)$
$BaTiO_3$ References 1, 2, 3	Tetragonal $C_{4v}(4mm)$	Cubic $O_h(m3m)$
$NaNO_2$ References 1, 3	Orthorhombic $C_{2v}(mm2)$	Orthorhombic $D_{2h}(mmm)$
Rochelle salt References 1, 2	Monoclinic $C_2(2)$	Orthorhombic $D_2(222)$

(1) Jona and Shirane;[15] (2) Burfoot;[16] (3) Abrahams and Keve;[17] some notations used in references 1, 2, 3 are rewritten so as to agree with those of Bradley and Cracknell.[18]

temperature at which the symmetry has been changed. However, the two critical points are identical for the second kind of ferroelectrics. The dielectric susceptibility χ for both kinds is defined by the Curie–Weiss law

TABLE 2.8.

General View on the Ferroelectric Phase Transitions

Ehrenfest's classification	Thermodynamic type	Dielectric nature	Critical points
First-order	Displacive	Induced dipoles	$T_0 < T_c$
Second-order	Order–disorder	Permanent dipoles	$T_0 = T_c$

$$\chi = \frac{C}{T - T_0} \tag{2.35}$$

with C being the Curie constant. In materials like barium titanate (induced-dipole type) the Curie constant is much greater than the temperature T_0 whereas in materials like KDP (permanent-dipole type) the Curie constant has the same order of magnitude as $T_0 = T_c$ (see Table 2.9).

TABLE 2.9.

Values of the Curie Constant

Substance	C (elstun.^{2} erg^{-1} cm^{-1} deg)	T_0, K	C/T_0
Barium titanate	12,000	393	30
KDP	259	122	2.1
Triglycine sulfate	255	322	0.79

The type of phase transition is closely related to the lattice symmetry through the dipole structure, on one hand, and through the number of formula units per unit cells, on the other hand. If the pseudo-spin one-half is associated with each dipole moment then the total pseudo-spin magnitude, S, is given by

$$S = (2^n - 1)/2 \tag{2.36}$$

with n being the total number of formula units for each unit cell. There is a simple rule connecting the type of phase transitions with the magnitude of S: displacive ferroelectrics are characterized by a small value of S whereas the order–disorder ferroelectrics are characterized by a large S compared to one (see Table 2.6).

A special attention should be paid to the induced-dipole phase transitions for which the two critical temperatures almost coincide, $T_0 = T_c$. (This is not the case with barium titanate where T_0 is observed well below T_c.) Here the two types of ferroelectrics are described by the same thermodynamic expansions. For this reason let us consider the general classification of phase transitions due to the change in the Gibbs free energy at the transition point.(19–21) If the free energy is a continuous function of temperature but the first derivative has a discontinuity at the transition point it is said that the system undergoes a first-order phase transition. However, if the free energy and the first derivative are both continuous but the second derivative has a discontinuity at this point it is said that the system undergoes a second-order phase transition. The changes in the Gibbs free energy, G, entropy, S, and volume, V, referring to the introduced definition are illustrated in Fig. 2.3 and Fig. 2.4.

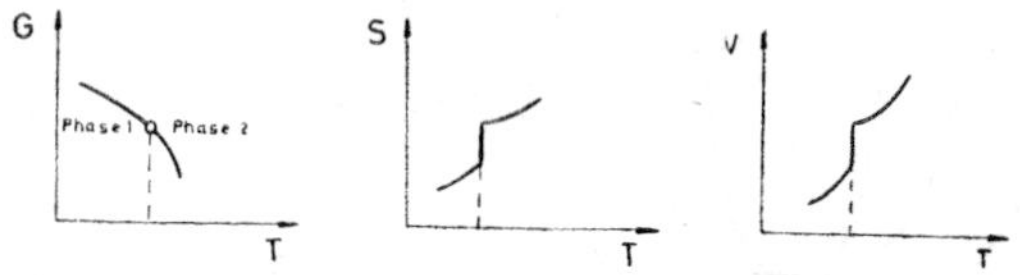

FIG. 2.3. Change of the Gibbs free energy, G, entropy, S, and volume, V, for a first-order phase transition.

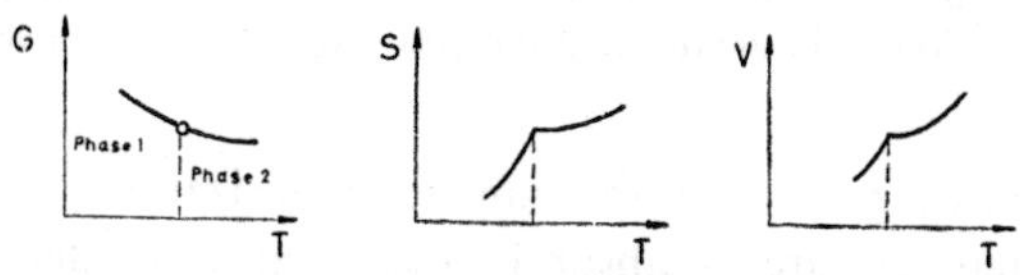

FIG. 2.4. Change of the Gibbs free energy, G, entropy, S, and volume, V, for a second-order phase transition.

Now we recall the well-known Landau–Lifshitz expansion of the free energy in terms of an order parameter, η, as follows(22, 23)

$$G_{(1)}-G_{(2)} = a\eta^2+b\eta^4+c\eta^6+\ \dots \tag{2.37}$$

where $a, b, c, \dots$ are temperature-dependent expansion parameters. Here the labeling (1) and (2) refers to the ordered and disordered phases, respectively. The order parameter η is usually taken as the polarization itself. There are two possibilities for the use of the thermodynamic coefficient in front of η^2, in particular,

Landau–Lifshitz expansion	General thermodynamic expansion
$a = -a_0(T_c - T)$	$a = -a_0(T_c - T)^{2\beta}$

with both a_0 and β being positive and temperature-independent quantities. A qualitative behavior of the free energy and polarization as temperature approaches the transition point from the ferroelectric side, $T \to T_c^-$, is illustrated in Fig. 2.5. On the basis of an empirical

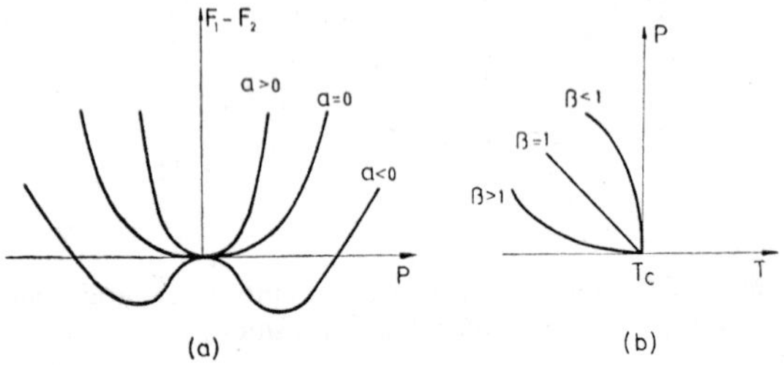

FIG. 2.5. Qualitative behavior of: (a) Helmholtz free energy, F, and (b) Spontaneous polarization, P, in the region just below the transition temperature.

evidence, showing that the spontaneous polarization just below the transition temperature is abruptly increased when temperature is decreased, we conclude that β has the value

$$0 < \beta < 1.$$

By a further analysis of the analytic property of the free energy as temperature approaches the transition point we arrive at the following conclusion: entropy of the system (first derivatives of G) has a finite discontinuity if the critical-point exponent β has the value 0.25, whereas specific heat of the system (second derivatives of G) has a finite discontinuity if β has the value 0.50. Therefore these two cases are associated with first-order and second-order phase transitions, respectively. The exact numerical results for the two mentioned cases are given by

$$\lim_{T\to T_c-} S_{(1)}-S_{(2)} = -a_0^2/(4b) \qquad (2.38a)$$

$$\lim_{T\to T_c-} C_{(1)}-C_{(2)} = a_0^2T_c/(2b) \qquad (2.38b)$$

where (1) and (2) refer to the ordered and disordered phases, respectively.

The problem of phase transition in a KDP structure does not seem to have been fully understood. On the one hand, the temperature dependence of the hysteresis loops and the dielectric constant of KDP(p) has been accurately determined in the neighborhood of the Curie temperature by Nazario and Gonzalo.(24) They measured the critical-point exponents β and δ appearing in the expansions

$$P \propto (T_c-T)^{\beta}, \qquad E \propto P^{\delta},$$

as T and E tend to the critical point and to zero, respectively, to obtain $\beta = 0.50\pm0.03$ and $\delta = 2.95\pm0.10$. Here E designates the magnitude of an applied electric field. The present data are consistent with a second-order phase transition. On the other hand, using dielectric and nuclear magnetic resonance data Craig(25) and Bjorkstam(26) suggested that the transition could be a first-order one accompanied by a small thermal hysteresis. Also by measuring thermodynamic parameters of this crystal in the critical region evidence is produced for a first-order phase transition with $\beta = 0.25$ by Strukov *et al.*(27)

Although the cited papers are in disagreement apparently one with another they are nevertheless perfectly consistent with the present analysis provided that the following fact is observed: in all hydrogen-bonded ferroelectrics the critical temperatures T_0 and T_c are equal to each other within a reasonable experimental error.

2.5. The Ising model with a transverse field: general theory

The first molecular theory of the phase transition in KDP was proposed by Slater.[28] Exactly as the diffraction experiment of Bacon and Pease[29] showed later Slater assumed that every proton occupies only one out of two possible positions along the oxygen–hydrogen–oxygen bonding line and out of four hydrogen bondings attached to every phosphate cluster (PO_4) he imagined himself only two protons to occupy the "close" positions all the time, while the other two being "away" from the cluster as demonstrated in Fig. 2.6. Disregarding

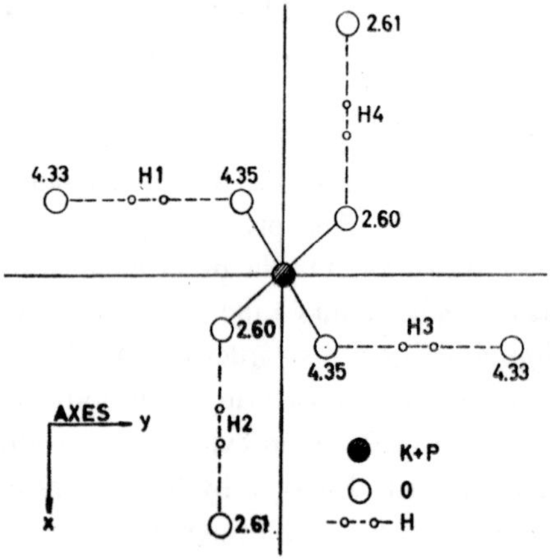

FIG. 2.6. Environment of a given central phosphate group (PO_4). The four nonequivalent proton sites are marked by H1, H2, H3, and H4. Numbers placed by the open circles designate the heights in units Å.

higher-order configurations Slater was able to explain the observed entropy change at the Curie temperature by predicting the first-order phase transition. By taking into account higher-order configurations as well Takagi[30] was able to show that the phase transition could be a second-order one. Although successful in appreciating many observed properties the Slater–Takagi model has a profound defect for it is lacking any explanation of the observed isotope effect. Not only KDP but also the whole family of related substances have this interesting feature as presented in Table 2.10.

TABLE 2.10.

Isotope Effect of Hydrogen-bonded Ferroelectrics and Antiferroelectrics (after Kanzig[14] and Jona and Shirane[15])

Substance	Curie temperature, K	Ratio T_{cd}/T_{cp}
	Ferroelectronics	
KH_2PO_4	122	
KD_2PO_4	223	1.83
KH_2AsO_4	96	
KD_2AsO_4	162	1.67
RbH_2AsO_4	111	
RbD_2AsO_4	178	1.62
RbH_2PO_4	148	
RbD_2PO_4	218	1.48
CsH_2AsO_4	143	
CsD_2AsO_4	212	1.48
CsH_2PO_4	159	
CsD_2PO_4	—	—
	Antiferroelectrics	
$NH_4H_2PO_4$	148	
$ND_4D_2PO_4$	242	1.64
$NH_4H_2AsO_4$	216	
$ND_4D_2AsO_4$	299	1.38

This feature indicates the important part played by the proton motion. Naturally a replacement of protons by deuterons and hence a change in the kinetic energy of the particles has a fundamental impact on the dynamical behavior of the whole lattice. There is presently a growing conviction that a specific proton–lattice interaction plays an essential part in the fundamental dynamics of hydrogen-bonded crystals. A systematic work to describe this phenomenon by assuming that protons perform a quantum-mechanical tunneling through a certain potential field is done by Cochran,[31] Blinc,[32] De Gennes,[33] Tokunaga and Matsubara,[34] as well as the present author.[35]

In what follows we shall formulate the Ising model with a transverse field ("tunneling" model). To do this we introduce the pseudo-spin one-half, in particular,

$$S_j^z = \tfrac{1}{2}, \qquad \text{or } -\tfrac{1}{2}$$

in order to specify the proton density concentration on the "right" or "left" equilibrium position, respectively. Here j labels the proton sites. The action of the pseudo-spin variable on the single-proton wave functions is defined by

$$\begin{aligned} S_j^z \Psi_{Rj}(y) &= \tfrac{1}{2}\, \Psi_{Rj}(y), \\ S_j^z \Psi_{Lj}(y) &= -\tfrac{1}{2}\, \Psi_{Lj}(y) \end{aligned} \tag{2.39}$$

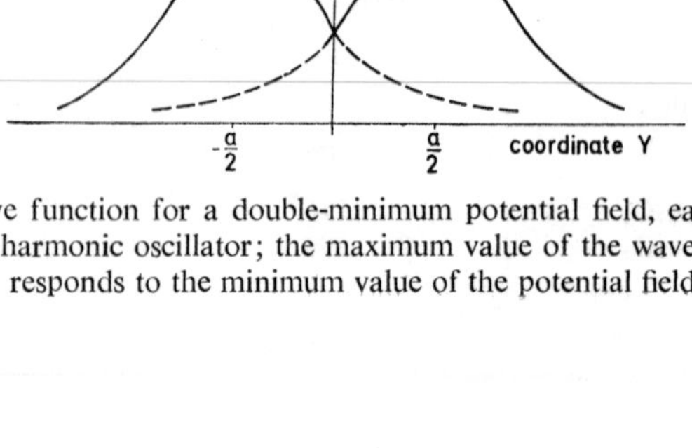

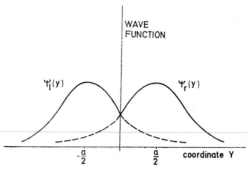

FIG. 2.7. Wave function for a double-minimum potential field, each field being identical to a harmonic oscillator; the maximum value of the wave function corresponds to the minimum value of the potential field.

where y designates the linear coordinate to coincide with the oxygen–hydrogen–oxygen line, as given in Fig. 2.7. Here a qualitative behavior of the wave function assuming a double-harmonic-oscillator potential well is presented with the two potential minima being separated by a distance of approximately 0.35 Å, the latter value has been derived from the neutron diffraction measurement of Bacon and Pease. Clearly at the transition temperature two remarkably distinct phases, one entirely ordered the other entirely disordered, exist together. Owing to the lattice symmetry imposed a symmetric shape for the double-minimum potential field is the only choice to be considered. Therefore the true eigenfunction representing the possible and yet still hypothetical proton states must be either symmetrical or antisymmetrical with respect to the substitution of the linear coordinate, y replacing -y. Hence

$$\begin{aligned}\Psi_{\text{sym}}(y) &= C_1[\Psi_{Rj}(y)+\Psi_{Lj}(y)],\\ \Psi_{\text{ant}}(y) &= C_2[\Psi_{Rj}(y)-\Psi_{Lj}(y)],\end{aligned} \tag{2.40}$$

with complete normalization factors. Any reasonable atomic model must be developed on the basis of a specific shape of the potential field, so the single-particle wave functions depend on this particular model. In general the unperturbed energy E_0 is thus separated into an energy doublet, one state corresponding to the function $\Psi_{\text{ant}}(y)$, the other one corresponding to the function $\Psi_{\text{sym}}(y)$ (see Fig. 2.8). The single-particle

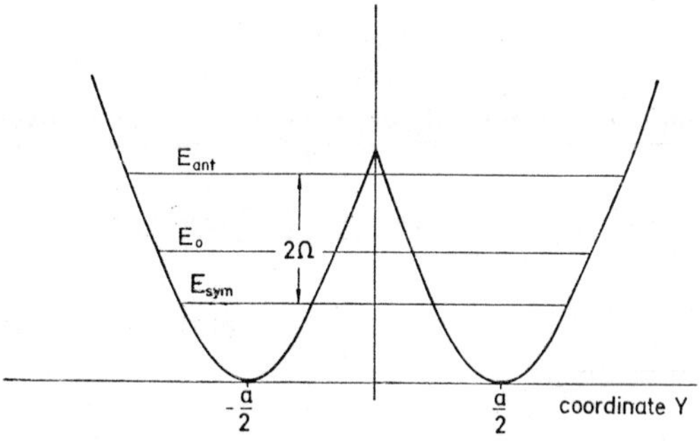

FIG. 2.8. Energy doublet as experienced by the particle whose motion is restricted by the double-minimum potential field.

hamiltonian as related to the motion in a double-minimum potential field,

$$\hat{h}(y) = -\frac{\hbar^2}{2m}\frac{d^2}{dy^2}V(y) \tag{2.41}$$

has the following matrix elements:

$$\begin{pmatrix} \langle R|\hat{h}|R\rangle & \langle R|\hat{h}|L\rangle \\ \langle L|\hat{h}|R\rangle & \langle L|\hat{h}|L\rangle \end{pmatrix} = E_0 I + 2\Omega S_j^x \tag{2.42}$$

where

$$\begin{aligned} |R\rangle &= \Psi_{Rj}(y), \\ |L\rangle &= \Psi_{Rj}(y), \end{aligned} \tag{2.43a}$$

$$\begin{aligned} E_0 &= \int \Psi_{Rj}^*(y)\,\hat{h}(y)\,\Psi_{Lj}(y)\,dy, \\ &= \int \Psi_{Lj}^*(y)\,\hat{h}(y)\,\Psi_{Lj}(y)\,dy \end{aligned} \tag{2.43b}$$

$$\Omega = (C_1^2 - C_2^2)E_0 - (C_1^2 + C_2^2)\int \Psi_{Rj}^*(y)\,\hat{h}(y)\,\Psi_{Lj}(y)\,dy. \tag{2.43c}$$

Here I designates a two-by-two unit matrix, S_j^x is the Pauli x matrix referring to the proton site j. Clearly the omega term appears as the kinetic energy of the particles thus having a neat quantum-mechanical origin.

Now all the contributions to the hamiltonian of the system are readily enumerated as follows:

$$\mathcal{H} = \mathcal{H}_0 + \mathcal{H}_1 + \mathcal{H}_2 \tag{2.44}$$

where the kinetic and potential energy of heavy ions is given by

$$\mathcal{H}_0 = \tfrac{1}{2}\sum_{f=1}^{N} p_f^* M_f^{-1} p_f + \tfrac{1}{2}\sum_{\langle fg\rangle} u_f^* K_{fg} u_g; \tag{2.45}$$

the proton kinetic energy ("transverse" field) and a proton–lattice interaction are given by

$$-\mathcal{H}_1 = 2\Omega\sum_{j=1}^{N} S_j^x + \tfrac{1}{2}\sum_{\langle jk;\,fg\rangle}(A_{fj}S_j^z u_f + A_{gk}^* S_k^z u_g^*); \tag{2.46}$$

and an interaction between the protons themselves ("dipole–dipole" forces) is given by

$$-\mathscr{H}_2 = \sum_{\langle jk \rangle} (B_{jk}S_j^x S_k^x + C_{jk}S_j^z S_k^z)$$
$$+ \sum_{\langle jkl \rangle} D_{jkl}S_j^z S_k^z S_l^z + \sum_{\langle jklm \rangle} F_{jklm}S_j^z S_k^z S_l^z S_m^z. \tag{2.47}$$

The first term in equation (2.47) describes a possible influence of the transverse field of one proton to the transverse field of another, whereas the remaining three terms take account of the long-range dipole–dipole forces (C terms) as well as of short-range forces having a three-body or even a four-body character (D and F terms). The three-body term actually disappears owing to the demand that this hamiltonian remains invariant with respect to the substitution of the pseudo-spin variable by its negative value at each proton site,

$$S_j^z \to -S_j^z,$$

leading to the condition $D_{jkl} = 0$. However, the four-body term may still survive in a general case of interest. It is worth noting that the total hamiltonian contains only those proton–proton interactions which are expressed by an even number of pseudo-spin variables, this observation being a straightforward consequence of the demand above introduced.

2.6. The Ising model with a transverse field: molecular-field approximation

It is clear that protons, or deuterons, in hydrogen-bonded ferroelectrics move under the influence of an internal (molecular) field which could be derived by a self-consistent averaging over the pseudo-spin components. Of course, such an averaging is justified in ferroelectric substances where a long-range dipole–dipole force dominates in comparison with the other interactions. There are presently two methods used to treat the proton–lattice interaction, one consists of an explicit account of it as done by Kobayashi[36] to obtain a correct mass-dependent transition temperature. The other method to be used here consists of eliminating the proton–lattice term on the cost of renormalizing the whole proton–proton interaction.

For this reason we transform the atomic displacement operator by introducing a pseudo-spin-dependent translation,

$$u_f = \xi_f + v_f \tag{2.48}$$

with v_f being a new displacement operator at a given lattice site. The translation ξ_f is to be determined by demanding that all linear terms in v_f must vanish identically. Therefore

$$\begin{aligned}\mathcal{H}_0+\mathcal{H}_1 = {} & \tfrac{1}{2}\sum_{f=1}^{N} p_f^* M_f^{-1} p_f + \tfrac{1}{2}\sum_{\langle fg\rangle} v_f^* K_{fg} v_g \\ & + \tfrac{1}{2}\sum_{\langle fg;\, jk\rangle} (v_f^* K_{fg}\xi_g + \xi_f^* K_{fg} v_g - A_{fj} S_j^z v_f - A_{gk}^* S_k^z v_g^*) \\ & - 2\Omega\sum_{j=1}^{N} S_j^x + \tfrac{1}{2}\sum_{\langle fg;\, jk\rangle} (\xi_f^* K_{fg}\xi_g - A_{fj} S_j^z \xi_f - A_{gk}^* S_k^z \xi_g^*).\end{aligned} \tag{2.49}$$

Here the first two sums represent the displaced harmonic oscillators, the third sum is required to vanish identically, whereas the fourth and fifth sums will be transformed accordingly. (A method used to separate the proton coordinates from those of heavy atoms is exposed in Appendix B.) By making the third sum vanish we arrive at a number of linear equations for the determination of the unknown quantities (ξ_f^* and ξ_g) in terms of the pseudo-spin variables (S_j^z and S_k^z). The result is thus given by determinants where every unknown quantity ξ_f^* depends linearly on a sum involving all the pseudo-spin variables. By returning to the last term in equation (2.49) one obtains a totally renormalized proton–proton interaction in the form

$$\sum_{\langle jk\rangle} \tilde{C}_{jk} S_j^z S_k^z$$

with $\tilde{C}_{jk}$ depending on the various coupling constants (K_{fg}, A_{fj} as well as A_{gk}^*).

Therefore the original interactions can be divided into the phonon and proton parts as follows:

$$\mathcal{H} = \mathcal{H}_{\text{ph}} + \mathcal{H}_{\text{pr}} \tag{2.50}$$

$$\mathcal{H}_{\text{ph}} = \tfrac{1}{2} \sum_{f=1}^{N} p_f^* M_f^{-1} p_f + \tfrac{1}{2} \sum_{\langle fg \rangle} v_f^* K_{fg} v_g,$$

$$-\mathcal{H}_{\text{pr}} = 2\Omega \sum_{j=1}^{N} S_j^x + \sum_{\langle jk \rangle} (B_{jk} S_j^x S_k^x + J_{jk} S_j^z S_k^z)$$

$$+ \sum_{\langle jklm \rangle} F_{jklm} S_j^z S_k^z S_l^z S_m^z, \tag{2.51}$$

where

$$J_{jk} = C_{jk} + \tilde{C}_{jk}.$$

One can observe that the proton part of the total hamiltonian is the Ising model with a transverse field.

Next we shall outline a self-consistent method to be used to obtain a reasonable estimation for the mass-dependent transition temperature. Let it be possible to make the two-body and four-body interactions linear by introducing a thermal average for every pseudo-spin operator with a certain fluctuation around the thermal average,

$$S_j^x = \langle S_x \rangle + \delta S_j^x,$$

$$S_j^z = \langle S_z \rangle + \delta S_j^z, \tag{2.52}$$

where δ has an obvious meaning. As a result of the above transformation there appears a molecular field with the components

$$H_x = 2\Omega + 2\zeta B \langle S_x \rangle,$$

$$H_z = 2\zeta J \langle S_z \rangle + 4\zeta F \langle S_z \rangle^3, \tag{2.53}$$

ζ being the number of nearest neighbors on an idealized cubic lattice. The purely proton hamiltonian is thus written in the first-order approximation at least,‡

$$\mathcal{H}_{\text{pr}} = -\sum_{j=1}^{N} (H_x \delta S_j^x + H_z \delta S_j^z) \tag{2.54}$$

with the sum running over all the proton sites. Therefore we may look at the total single-particle pseudo-spin, S_j, with the components S_j^x and S_j^z as if it moves under the influence of a self-consistent molecular field, H, with the components H_x and H_z. This is illustrated in Fig. 2.9.

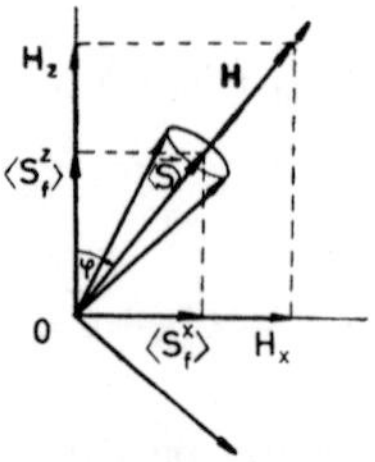

FIG. 2.9. General self-consistent molecular field, $\boldsymbol{H}$, acting on a given pseudo-spin, $\boldsymbol{S}$.

Now the introduced thermal average $\langle S_z \rangle$ can be estimated as follows:

$$\begin{aligned} \langle S_z \rangle &= Z^{-1} \operatorname{Tr} S_j^z \exp(-\beta \mathcal{H}), \\ Z &= \operatorname{Tr} \exp(-\beta \mathcal{H}), \\ \beta &= 1/(kT). \end{aligned} \tag{2.55}$$

A similar definition holds also for the other thermal average, $\langle S_x \rangle$. By the inspection of Fig. 2.9 one writes

$$\frac{\langle S_z \rangle}{H_z} = \frac{\langle S_{z'} \rangle}{H} \tag{2.56}$$

‡ In order to evaluate the introduced thermal averages we may use the following self-consistent molecular field equation in place of equation (2.51),

$$\mathcal{H}_{\text{pr}} \cong -\sum_{j=1}^{N} \boldsymbol{H} \cdot \boldsymbol{S}_j = -\sum_{j=1}^{N} (H_x S_j^x + H_z S_j^z)$$

the two expressions being equivalent to each other from the viewpoint of statistical mechanics.

where H designates the total molecular field. For spin one-half

$$\langle S_{z'} \rangle = \frac{1}{2} \tanh \left(\frac{\beta H}{2} \right) \tag{2.57}$$

with H replacing $\mathcal{H}$ in equation (2.55). Using equations (2.53), (2.56), and (2.57) we arrive at the following self-consistent molecular field equation:

$$H = \zeta [J + 2F \langle S_z \rangle^2] \tanh \left(\frac{\beta H}{2} \right). \tag{2.58}$$

It is clear that the magnitude of the molecular field is a temperature-dependent quantity as one can observe from Table 2.11. Here are

$$\begin{aligned} H &= \zeta [J + 2F \langle S_z \rangle^2]\, Y, \\ 2kT &= \zeta [J + 2F \langle S_z \rangle^2]\, \Theta, \\ Y &= \tanh (Y/\Theta) \end{aligned} \tag{2.59}$$

and

$$Y \cong \sqrt{(1 - \Theta^2)}. \tag{2.60}$$

TABLE 2.11.

Numerical Analysis of the Molecular-field Equations

Θ	Y, equation (2.59) exact result	Y, equation (2.60) approximation
0	1	1
0.1	1	0.995
0.2	1	0.980
0.3	0.997	0.954
0.4	0.983	0.917
0.5	0.958	0.866
0.6	0.907	0.8
0.7	0.829	0.714
0.8	0.710	0.6
0.85	0.630	0.527
0.9	0.525	0.436
0.95	0.380	0.312
1	0	0

The latter equation, (2.60), is a very good approximation to the original molecular-field equation (2.59), as one can observe from the presented numerical analysis (Table 2.11).

At the transition temperature one can write

$$\beta = \beta_c, \quad \langle S_z \rangle = 0,$$

leading to

$$H = H_x = \zeta J \tanh(\beta_c H_x/2). \tag{2.61}$$

Since the x component of the molecular field, H_x, if related to the deuterons is smaller than that related to the protons by a factor

$$\sqrt{(m_d/m_p)} = \sqrt{2}$$

then an isotope effect on the transition temperature is immediately seen (Fig. 2.10). In a limiting case where this component tends to zero we discover

$$kT_{c,0} = \zeta J/2. \tag{2.62}$$

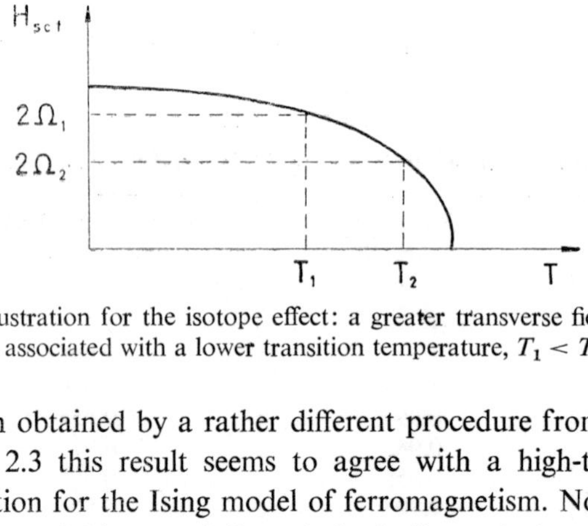

FIG. 2.10. Illustration for the isotope effect: a greater transverse field, $\Omega_1 > \Omega_2$, is associated with a lower transition temperature, $T_1 < T_2$.

Although obtained by a rather different procedure from that used in Section 2.3 this result seems to agree with a high-temperature approximation for the Ising model of ferromagnetism. No doubt the "tunneling" model is essentially and physically equivalent to the Ising model within a self-consistent molecular field approximation at least.

The present model can also supply a satisfactory explanation for the isotope effect on the saturation polarization, P_s. By definition

$$P_s = N\mu\langle\sigma_{\text{prot}}\rangle \tag{2.63}$$

where N and μ designate the total number of electric dipoles per unit volume and the dipole moment of a complex $K^+(H_2PO_4)^-$, respectively. Here is

$$\langle\sigma_{\text{prot}}\rangle = 2\langle S_z\rangle = 2\cos\Phi\langle S_{z'}\rangle = \cos\Phi\sqrt{(1-\Theta^2)} \qquad (2.64)$$

with

$$\cos\Phi = \sqrt{(1-\sin^2\Phi)} = \sqrt{\{1-(H_x/H)^2\}}. \qquad (2.65a)$$

It is clear that the rotation angle in the pseudo-spin space Φ is a function of temperature so that $\cos\Phi$ vanishes at the transition temperature and has a maximum value at the absolute zero,

$$(\cos\Phi)_{\max} = \sqrt{\{1-(H_{x,0}/H_0)^2\}}, \qquad T = 0, \qquad (2.65b)$$

where

$$H_0 = H(T=0) = \zeta[J+2F\langle S_z\rangle^2_{(0)}]$$

by equation (2.59) and Table 2.11. Here $H_{x,0}$ designates the zero-temperature value of the transverse field. These two equations are determined as follows.

First, it is a reasonable assumption to neglect the four-body interactions introduced in the previous section by letting

$$F = 0$$

in all molecular-field equations. This leads to

$$H_0 = \zeta J = H_x \coth\left(\frac{H_x}{2kT_c}\right) \qquad (2.66)$$

by equation (2.61).

Second, it is reasonable to assume that Nernst's theorem[22] operates by which "the entropy of any body vanishes at absolute zero". This means that any part of the system must be in the ground state so as to enable the system to achieve a perfect order. In particular,

$$(\cos\Phi)_{\max} = 1, \quad H_{x,0} = 0, \qquad (2.67)$$

at absolute zero. A low-temperature expansion is identical to the self-consistent molecular field

$$\cos \Phi = \tanh\left(\frac{H_0}{2kT}\right) \tag{2.68}$$

by equation (2.58).

If the approximation (2.68) is extended over the whole ferroelectric region including the transition temperature then one can write

$$\frac{H_0}{2kT} = \frac{Y_0}{\Theta}$$

$$\begin{aligned} \cos \Phi &= \tanh (Y_0/\Theta) = Y_0 \\ &= \sqrt{\{1-(T/T_c)^2\}} \end{aligned} \tag{2.69}$$

where we use equations (2.59) and (2.60). Therefore the order parameter of the proton system is given by

$$\langle \sigma_{\text{prot}} \rangle = \sqrt{\{1-(T/T_c)^2\}}\sqrt{(1-\Theta^2)} \tag{2.70}$$

which vanishes at the transition temperature and has a maximum value at absolute zero as it should be. A low-temperature expansion has the form

$$\langle \sigma_{\text{prot}} \rangle = 1-\sigma_1(kT/H_0)^2 + \ldots \tag{2.71a}$$

with

$$\sigma_1 = \frac{1}{2}\left[\left(\frac{H_0}{kT_c}\right)^2 + 4\right]. \tag{2.71b}$$

It can be observed from Table 2.12 that the zero-temperature molecular field, H_0, is almost a mass-independent quantity whereas the expansion coefficient σ_1 is highly mass dependent in such a way that the order parameter, $\langle \sigma_{\text{prot}} \rangle$, decreases with temperature more slowly as the mass of the particles is increased. This seems to agree with a recent work of Azoulay *et al.*[39] who measured the ratio of the two spontaneous polarizations with the result

$$P_s(KD_2PO_4)/P_s(KH_2PO_4) = 1.2.$$

TABLE 2.12.

High-temperature Transverse Field (H_x), Zero-temperature Molecular Field (H_0), and Expansion Coefficient (σ_1) for Hydrogen-bonded Ferroelectrics

Substance	T_c, K	H_x, cm^{-1}	H_0, cm^{-1}	σ_1
KH_2PO_4	122	374‡ 345§	383‖ 357 ¶	12.20‖ 10.86 ¶
KD_2PO_4	223	261‡ 220§	380‖ 360 ¶	5.00‖ 4.70 ¶

‡ Identified with the single-particle low-lying energy difference across a double-minimum potential field.(37)

§ Identified with the single-particle strong line difference as observed by a laser-excited Raman spectrum.(38)

‖ Using the former set of data,

¶ Using the latter set of data.

2.7. The critical-point exponents

If the system of ferromagnetic (ferroelectric) spins is subject to an external magnetic (electric) field then $T = T_c$ becomes a singular point thus making inappropriate the Landau–Lifshitz expansion in terms of an order parameter. Not only the magnetization (polarization) but also the magnetic (dielectric) susceptibility, specific heat, or any other characteristic quantity, are no longer determined by this kind of expansion. These quantities are represented in a contemporary research by the critical-point exponents which are introduced as follows:(40–43)

$M(P) \propto (T_c - T)^\beta$ below T_c,

$H \propto M^\delta$ both below

$E \propto P^\delta$ and above T_c,

$$\begin{aligned} \chi &\propto (T-T_c)^{-\gamma} && \text{above } T_c, \\ \chi &\propto (T_c-T)^{-\gamma'} && \text{below } T_c, \\ C_H(C_E) &\propto (T-T_c)^{-\alpha} && \text{above } T_c, \\ C_H(C_E) &\propto (T_c-T)^{-\alpha'} && \text{below } T_c, \end{aligned} \tag{2.72}$$

to mention very few of them. In all listed cases the external field tends to zero. Numerical values obtained by various theories are collected in Table 2.13.

TABLE 2.13.

Critical-point Exponents in Various Theories

Physical quantity	Exponent	Landau theory	Two-dimensional Ising model	Three-dimensional Ising model
Order parameter	β	$\frac{1}{2}$	$\frac{1}{8}$	0.313 ± 0.004
Applied field	δ	3	15	5.20 ± 0.15
Suscepti-bility	γ	1	$\frac{7}{4}$	1.250 ± 0.001
Susceptibility	γ'	1	$\frac{7}{4}$	1.310 ± 0.050
Specific heat	α	0‡	0§	$0.0 \leq \alpha \leq 0.25$
Specific heat	α'	0‡	0§	$0.066^{+0.16}_{-0.04}$
Observed in:		Ferro-electric phase transi-tions		Ferromagnetic phase transitions

‡ There is a discontinuity in the specific heat as the applied field tends to zero.
§ There is a logarithmic divergence in the specific heat as the applied field tends to zero.

To establish a possible interdependence for the critical-point exponents we shall employ well-known thermodynamic relations. For this reason we recall a certain analogy existing between pressure versus volume and magnetic (electric) field versus magnetization (polarization).[40] Having designated the specific heats at constant field and at constant magnetization (polarization) with $C_H(C_E)$ and $C_M(C_P)$, respectively, and using the above analogy of Yang and Lee[40] we can write

$$C_H = C_M + T\left(\frac{\partial M}{\partial T}\right)_H^2 \left(\frac{\partial H}{\partial M}\right)_T \tag{2.73}$$

where the meaning of the used symbols is clear.

The stability condition for the Gibbs free energy states that the specific heat at constant magnetization is not negative,[44]

$$C_M = T\left(\frac{\partial S}{\partial T}\right)_M = -T\left(\frac{\partial^2 G}{\partial T^2}\right)_M \geqslant 0 \tag{2.74}$$

where

$$G = F + MH,$$

F designates the Helmholtz free energy. Using equations (2.73) and (2.74) we obtain

$$C_H \geqslant T\left(\frac{\partial M}{\partial T}\right)_H^2 \left(\frac{\partial H}{\partial M}\right)_T \tag{2.75}$$

to be applied to the region just below the transition temperature as H tends to zero. The behavior of the second factor in (2.75) is identical to that of an inverse susceptibility,

$$\chi^{-1} = \left(\frac{\partial H}{\partial M}\right)_T \propto (T_c - T)^{\gamma'};$$

therefore

$$C_H \propto (T_c - T).^{2(\beta-1)+\gamma'}. \tag{2.76}$$

Having compared equations (2.72) with the behavior (2.76) one arrives at the inequality

$$\alpha' + 2\beta + \gamma' \geqslant 2. \tag{2.77}$$

On the other hand, one can write

$$C_H \geqslant T\left(\frac{\partial M}{\partial T}\right)_H\left(\frac{\partial H}{\partial M}\right)_T\left(\frac{\partial M}{\partial T}\right)_H$$

$$= T\left(\frac{\partial M}{\partial T}\right)_H\left(\frac{\partial H}{\partial T}\right) \propto (T_c - T)^{\beta(1+\delta)-2} \tag{2.78}$$

where the following behavior was used:

$$H \propto (T_c - T)^{\beta\delta}.$$

Having compared equations (2.72) with the behavior (2.78) one arrives at another inequality

$$\alpha' + \beta(1+\delta) \geqslant 2. \tag{2.79}$$

It is expected that all cluster models predict certain symmetry relations among the critical-point exponents;[44] for instance, the susceptibility and specific heat just above and below the transition temperature should be expressed by similar expansions. Therefore we may expect

$$\gamma' = \gamma, \qquad \alpha' = \alpha. \tag{2.80}$$

The interdependent relations (2.77), (2.79), and (2.80) in a contemporary research are named the scaling-law relations.

Equations (2.80) are one of the straightforward consequences of the pseudo-spin method. Indeed, since the total hamiltonian is invariant under the replacement of each pseudo-spin by its negative, $\boldsymbol{S}_j \to -\boldsymbol{S}_j$, and simultaneously by $H \to -H$, it follows at once that the relevant thermodynamic expressions are equivalent to each other in the two limiting temperature regions around the critical point. (Here an external pressure according to Yang and Lee[40] is replaced by a magnetic field so as to ensure the total hamiltonian become invariant under the replacement $p \to -p$.) Therefore the relevant temperature expansion in terms of $T_c - T$, or $T - T_c$, must contain the same critical-point exponent. Hence the equations (2.80).

References

1. D. Wagner *Introduction to the Theory of Magnetism*, Pergamon Press, Oxford, 1972.
2. R. M. Bozorth, in: *American Institute of Physics Handbook*, McGraw-Hill Book Co., Inc., New York, 1957.
3. I. Ursu, *La Résonance Paramagnétique Électronique*, DUNOD, Paris, 1968.
4. E. Ising, *Z. Physik* **31,** 253 (1925).
5. L. Onsager, *Phys. Rev.* **65,** 117 (1944).
6. C. N. Yang, *Phys. Rev.* **85,** 808 (1952).
7. K. Huang, *Statistical Mechanics*, John Wiley & Sons, Inc., New York, 1963.
8. W. Heisenberg, *Z. Physik* **49,** 619 (1928).
9. D. D. Betts, C. J. Elliott, and M. H. Lee, *Can. J. Phys.* **48,** 1566 (1970).
10. J. H. Van Vleck, *J. Chem. Phys.* **9,** 85 (1941).
11. J. S. Smart, *Phys. Rev.* **86,** 968 (1952).
12. L. Néel, *Ann. Phys.* (Paris) **3,** 137 (1948).
13. H. D. Megaw, *Ferroelectricity in Crystals*, Methuen & Co. Ltd., London, 1957.
14. W. Kanzig, Ferroelectrics and antiferroelectrics, in: *Solid State Physics*, Vol. 4, ed. by F. Seitz and D. Turnbull, Academic Press, New York, 1957.
15. F. Jona and G. Shirane, *Ferroelectric Crystals*, Pergamon Press, Oxford, 1962.
16. J. C. Burfoot, *Ferroelectrics*, D. Van Nostrand Co. Ltd., London, 1967.
17. S. C. Abrahams and E. T. Keve, *Ferroelectrics* **2,** 129 (1971).
18. C. J. Bradley and A. P. Cracknell, *The Mathematical Theory of Symmetry in Solids*, Clarendon Press, Oxford, 1972.
19. P. Ehrenfest, *Proc. Kon. Ned. Akad. Wet.* (Amsterdam) **36,** 153 (1933).
20. D. Ter Haar, *Elements of Statistical Mechanics*, chap. IX, Constable & Co., London, 1955.
21. A. B. Pippard, *The Elements of Classical Thermodynamics*, chap. 9, Cambridge University Press, London, 1964.
22. L. D. Landau and E. M. Lifshitz, *Statistical Physics*, chap. XIV, Pergamon Press, Oxford, 1959.
23. A. J. Dekker, *Solid State Physics*, sections 8–7, MacMillan & Co. Ltd., London, 1962.
24. I. Nazario and J. A. Gonzalo, *Solid State Commun.* **7,** 1305 (1969).
25. P. P. Craig, *Phys. Lett.* **1,** 140 (1966).
26. J. L. Bjorkstam, *Phys. Rev.* **153,** 599 (1967).
27. B. A. Strukov, M. A. Korzhuev, A. Baddur, and V. A. Koptsik, *Fiz. Tverd. Tela* **13,** 1872 (1971) [English trans.: *Soviet Phys. — Solid State* **13,** 1569 (1972)].
28. J. C. Slater, *J. Chem. Phys.* **9,** 16 (1941).
29. G. E. Bacon and R. S. Pease, *Proc. Roy. Soc.* (London) A **230,** 359 (1955).

30. Y. Takagi, *J. Phys. Soc. Japan* **3**, 271 (1948).
31. W. Cochran, *Adv. Phys. (Phil. Mag. Suppl.)* **9**, 387 (1960); *ibid.* **10**, 401 (1961).
32. R. Blinc, *J. Phys. Chem. Solids* **13**, 204 (1960).
33. P. G. De Gennes, *Solid State Commun.* **1**, 132 (1963).
34. M. Tokunaga and T. Matsubara, *Prog. Theor. Phys.* **35**, 581 (1966); M. Tokunaga, *ibid.* **36**, 857 (1966).
35. L. Novaković, *J. Phys. Chem. Solids* **27**, 1469 (1966).
36. K. K. Kobayashi, *J. Phys. Soc. Japan* **24**, 497 (1968).
37. L. Novaković, *J. Phys. Chem. Solids* **31**, 431 (1970).
38. I. P. Kaminow, R. C. C. Leite, and S. P. S. Porto, *J. Phys. Chem. Solids* **36**, 2085 (1965).
39. J. Azoulay, Y. Grinberg, I. Pelah, and E. Wiener, *J. Phys. Chem. Solids* **29**, 843 (1968).
40. C. N. Yang and T. D. Lee, *Phys. Rev.* **87**, 404, 410 (1952).
41. M. E. Fisher, *Rep. Prog. Phys.* **30**, 615 (1967); M. E. Fisher in: *Contemporary Physics*, Vol. I, Trieste Symposium 1968, International Atomic Energy Agency, Vienna, 1969.
42. L. P. Kadanoff *et al., Rev. Mod. Phys.* **39**, 395 (1967).
43. H. E. Stanley, *Introduction to Phase Transitions and Critical Phenomena*, Oxford University Press, London, 1971.
44. P. W. Kasteleyn in: *Fundamental Problems in Statistical Mechanics*, II, ed. by E. G. D. Cohen, North-Holland Publishing Co., Amsterdam, 1968.

CHAPTER 3

Magnetic Elementary Excitations

3.1. The second quantization method

In a quantum-mechanical investigation of the system consisting of a large number of identical particles which interact in an arbitrary manner we often use the second quantization method. This method is particularly useful in a system where the number of particles is a variable quantity. We shall at first expose the essential notions about the wave functions which describe the system of identical particles.

Consider a system consisting of N identical noninteracting particles, e.g. the free electrons in a metal, or the phonons in a crystal. The Schrödinger equation for a stationary state in this system may be written

$$\sum_{i=1}^{N}\left[-\frac{\hbar^2}{2m}\Delta_i+V(\boldsymbol{r}_i)\right]\Psi(\boldsymbol{r}_1, \boldsymbol{r}_2, \ldots, \boldsymbol{r}_N) = E\Psi(\boldsymbol{r}_1, \boldsymbol{r}_2, \ldots \boldsymbol{r}_N) \tag{3.1}$$

where the first term refers to the kinetic energy in an operation form,

$$\Delta_i = \frac{\partial^2}{\partial x_i^2}+\frac{\partial^2}{\partial y_i^2}+\frac{\partial^2}{\partial z_i^2}, \tag{3.2}$$

m is the mass of the particles. The second term represents the potential energy as a function of the position vectors $\boldsymbol{r}_i = \{x_i, y_i, z_i\}$. E denotes the binding energy of the whole system in a given stationary state. The solution to the above equation is usually taken in the form

$$\Psi = \psi_{q_1}(\mathbf{r}_1)\,\psi_{q_2}(\mathbf{r}_2)\,\ldots\,\psi_{q_N}(\mathbf{r}_N) \tag{3.3}$$

where q_i labels a set of quantum numbers characterizing a given stationary state. Every q_i represents a full set of quantum numbers which describe the state as occupied by an individual particle. The functions ψ_{qi} are the solutions to the Schrödinger equation for one particle,

$$\left[-\frac{\hbar^2}{2m}\Delta_i + V(\mathbf{r}_i)\right]\psi_{q_i}(\mathbf{r}_i) = E_{q_i}\psi_{q_i}(\mathbf{r}_i). \tag{3.4}$$

However, the wave function (3.3) does not satisfy the symmetry requirement. In general it is neither symmetrical nor antisymmetrical with respect to the exchange of the coordinates of any two particles. Since equation (3.1) is a linear combination of the solution (3.3) is also a solution. Therefore a general solution which is properly symmetrized or antisymmetrized must be a superposition of the elementary solutions of the type (3.3).

Consider, for instance, a system of two identical particles. Clearly the possible wave functions are given by a combination which is either symmetric or antisymmetric,

$$\Psi_{s,a} = \frac{1}{\sqrt{2}}\,[\psi_1(\mathbf{r}_1)\,\psi_2(\mathbf{r}_2) \pm \psi_2(\mathbf{r}_1)\,\psi_1(\mathbf{r}_2)], \tag{3.5}$$

where s and a correspond to $+$ and $-$, respectively. Each wave function is normalized to unity.

The above result can be generalized to a system having an arbitrarily large number of identical particles, N. In this case we require for the total wave function to be either symmetrical or antisymmetrical with respect to the permutation of the coordinates belonging to any pair of particles. In the former case we say that the particles obey Bose–Einstein statistics, whereas in the latter case we say that the particles obey Fermi–Dirac statistics. The former particles are called bosons which according to a Pauli's investigation(1–2) have an integer spin in units $\hbar$ (e.g. photons, phonons, π-mesons). The latter particles are called fermions which according to the same Pauli's investigation have a half-integer spin in units $\hbar$ (e.g. electrons, nucleons, ferroelectric protons or deuterons).

To expose the problem of second quantization in a complete form we start from the assumption that the system of N noninteracting particles, say bosons, is submitted to an external field. Then every boson occupies one state which belongs to the set of states whose energies are E_0, E_1, E_2, ... Let us denote the corresponding wave functions with $\psi_{q_0}(\boldsymbol{r})$, $\psi_{q_1}(\boldsymbol{r})$, $\psi_{q_2}(\boldsymbol{r})$, ... Clearly the energy of such a general state is given by the matrix element of the single particle operator $\hat{H}_1(\boldsymbol{r})$,

$$E_i = \int_{-\infty}^{\infty} \psi^*_{q_i}(\boldsymbol{r}_1)\, \hat{H}_1(\boldsymbol{r}_1)\, \psi_{q_i}(\boldsymbol{r}_1)\, d\tau_1, \tag{3.6}$$

whereas the interaction between the particles is given by the matrix element of the two-particle operator

$$V_{ij} = \int d\tau_1 \int d\tau_2 \psi^*_{q_i}(\boldsymbol{r}_1)\, \psi^*_{q_j}(\boldsymbol{r}_2)\, \hat{H}_{12}(\boldsymbol{r}_1, \boldsymbol{r}_2)\, \psi_{q_j}(\boldsymbol{r}_2)\, \psi_{q_i}(\boldsymbol{r}_1), \tag{3.7}$$

$$W_{ij} = \int d\tau_1 \int d\tau_2 \psi^*_{q_i}(\boldsymbol{r}_1)\, \psi^*_{qi}(\boldsymbol{r}_2)\, \hat{H}_{12}(\boldsymbol{r}_1, \boldsymbol{r}_2)\, \psi_{q_i}(\boldsymbol{r}_2)\, \psi_{q_j}(\boldsymbol{r}_1). \tag{3.8}$$

We will in the following name the quantities V_{ij} and W_{ij} the direct and exchange integrals, respectively.

Let us assume that there are n_i identical bosons in a given state with the energy E_i. Clearly we must have $\Sigma n_i = N$, where the sum runs over all allowed states of the system. Since the Pauli exclusion principle does not apply to the boson system, any individual number n_i may be as large as possible provided that the total number of bosons N is sufficiently large. Therefore, instead of characterizing the system by giving the set of energies and the corresponding wave functions we may characterize it by giving the set of energies and the corresponding occupation numbers, n_i. Since the system is subject to an external field the particles are allowed in the course of time to change their positions in the configurational space and hence to make the transitions from one energy state to another. It is therefore desirable to introduce a specific operational formalism in order to account for the dynamical behavior of the particles. For this purpose we consider the individual wave functions as the amplitudes of a field the absolute magnitude of which determines the probability of finding the particle in the state defined by a particular individual wave function. Hence we replace the wave functions $\psi_{q_i}(\boldsymbol{r})$ and $\psi^*_{q_i}(\boldsymbol{r})$ by a set of annihilation and creation operators as follows:

$$\begin{aligned} \psi_{q_i}(\boldsymbol{r}) &\to N^{1/2}\psi_{q_i}(\boldsymbol{r})\, a_i, \\ \psi^*_{q_i}(\boldsymbol{r}) &\to N^{1/2}\psi^*_{q_i}(\boldsymbol{r})\, a_i^\dagger, \end{aligned} \tag{3.9}$$

where the operator a_i takes (annihilates) a boson from the state E_i to another state, and the operator $a_i^\dagger$ brings (creates) to the state E_i a boson from another state.

The above operators satisfy the following commutation relations:

$$\begin{aligned} [a_i, a_i^\dagger]_{(-)} &= \Delta(i-j), \\ [a_i, a_j]_{(-)} &= [a_i^\dagger, a_i^\dagger]_{(-)} = 0 \end{aligned} \tag{3.10}$$

where the meaning of $(-)$ is obvious. The second relation embodies the fact that any two bosons may be interchanged without changing the sign of the wave function. To find a physical meaning for the first commutation relation in equation (3.10) we introduce the state vectors which characterize the given state completely. These state vectors are designated as

$$\Phi_{q_0q_1\ldots q_i\ldots}(n_0, n_1, n_2, \ldots) \equiv |\, n_0, n_1, n_2, \ldots\rangle \tag{3.11}$$

to indicate that there are n_i identical bosons in a given state E_i. Very often the shorthand notation of the right side of equation (3.11) is used. Also we introduce the hermitian conjugate to the above state vectors as

$$\Phi^*_{q_0q_1\ldots q_i\ldots}(n_0, n_1, n_2, \ldots) \equiv \langle n_0, n_1, n_2, \ldots |. \tag{3.12}$$

The matrix elements for the creation and annihilation operators are most conveniently defined by

$$\begin{aligned} a_i^\dagger\,|\, n_0, n_1, \ldots n_i, \ldots\rangle &= \sqrt{(n_i+1)}\,|\, n_0, n_1, \ldots n_i+1, \ldots\rangle, \\ a_i\,|\, n_0, n_1, \ldots n_i, \ldots\rangle &= \sqrt{(n_i)}\,|\, n_0, n_1, \ldots n_i-1, \ldots\rangle. \end{aligned} \tag{3.13}$$

We may observe that the occupation number operator

$$N_i = a_i^\dagger a_i \tag{3.14}$$

has the following matrix element:

$$N_i \mid n_0, n_1, \ldots n_i, \ldots\rangle = n_i \mid n_0, n_1, \ldots n_i, \ldots\rangle. \tag{3.15}$$

Now, the first commutation relation (3.10) has a set of equivalent operational equations in the form

$$\begin{aligned} a_i^\dagger N_i &= (N_i - 1)\, a_i^\dagger, \\ a_i^{\dagger 2} a_i^2 &= a_i^\dagger N_i a_i = N_i (N_i - 1), \\ a_i^{\dagger 3} a_i^3 &= a_i^{\dagger 2} N_i a_i^2 = N_i (N_i - 1)(N_i - 2), \\ &\cdot \quad \cdot \quad \cdot \quad \cdot \quad \cdot \\ a_i^{\dagger k} a_i^k &= a_i^{\dagger (k-1)} N_i a_i^{(k-1)} = N_i (N_i - 1) \ldots (N_i - k + 1). \end{aligned} \tag{3.16}$$

Therefore the matrix elements of the above operational identities become

$$\begin{aligned} &\langle n_0, n_1, \ldots n_i, \ldots \mid a_i^{\dagger k} a_i^k \mid n_0, n_1, \ldots n_i, \ldots\rangle \\ &= n_i (n_i - 1) \ldots (n_i - k + 1) < n_0, n_1, \ldots n_i \ldots \mid n_0, n_1, \ldots n_i \ldots\rangle \end{aligned} \tag{3.17}$$

Clearly the operator $a_i^{\dagger k}\ a_i^k$ is a product of some operator and its hermitian conjugate, so equation (3.17) cannot be negative. Hence, if n_i is not an integer we might find k such that

$$k > n_i + 1,$$

so equation (3.17) would become negative which is a contradiction. Therefore every n_i is a positive integer.

Using equations (3.6–3.10) we may write the total hamiltonian in the form

$$\mathcal{H}_{SQ} = \sum_i E_i a_i^\dagger a_i + \frac{N}{2(N-1)} \sum_{i \neq j} (V_{ij} + W_{ij})\, a_i^\dagger a_j^\dagger a_j a_i, \tag{3.18}$$

where E_i, V_{ij}, and W_{ij} denote the matrix elements of the single-particle and two-particle operators, respectively. The summations are taken over all allowed states, so if N is sufficiently large then $N/(N-1)$ may be replaced by 1.

The exposed equations constitute the basis for the second quantization method. Although formally developed for boson systems this method may equally be applied to fermion systems provided that some necessary modifications are made.

Indeed, let us assume that $\psi_{q_i}(\boldsymbol{r})$ and $\psi^*_{q_i}(\boldsymbol{r})$ are the single-particle wave functions of a system of fermions. Here we introduce a substitution similar to that in equation (3.9),

$$\begin{aligned} \psi_{q_i}(\boldsymbol{r}) &\to N^{1/2}\psi_{q_i}(\boldsymbol{r})\, b_i, \\ \psi^*_{q_i}(\boldsymbol{r}) &\to N^{1/2}\psi^*_{q_i}(\boldsymbol{r})\, b_i^\dagger, \end{aligned} \tag{3.19}$$

where the operators b_i and $b_i^\dagger$ annihilate or create the fermion at the state with the energy E_i. These operators satisfy the following anti-commutation relations:

$$\begin{aligned} [b_i, b_j^\dagger]_{(+)} &= \Delta(i-j), \\ [b_i, b_j]_{(+)} &= [b_i^\dagger, b_j^\dagger]_{(+)} = 0 \end{aligned} \tag{3.20a}$$

where the meaning of $(+)$ is obvious. If $i = j$ then the second relation becomes

$$b_j^2 = b_j^{\dagger 2} = 0, \tag{3.20b}$$

i.e. a twofold application of either the creation or annihilation operator gives always zero.

Having introduced the occupation number operator

$$N_i = b_i^\dagger b_i \tag{3.21}$$

there follow the relations

$$\begin{aligned} b_i N_i &= (1-N_i)\, b_i \\ N_i^2 = b_i^\dagger b_i N_i &= b_i^\dagger (1-N_i)\, b_i = N_i. \end{aligned} \tag{3.22}$$

Hence, if n_i denotes the number of fermions allowed to the state E_i then

$$n_i = 0 \quad \text{or} \quad 1. \tag{3.23}$$

This is in agreement with the Pauli exclusion principle.

Now it is easy to see that an equivalent equation to equation (3.18) using the second quantization method may be written in the form

$$\mathcal{H}_{SQ} = \sum_i E_i b_i^\dagger b_i + \frac{N}{2(N-1)} \sum_{i \neq j} (V_{ij} - W_{ij})\, b_i^\dagger b_j^\dagger b_j b_i \qquad (3.24)$$

where the quantities E_i, V_{ij}, and W_{ij} denote the matrix elements of the single-particle and two-particle operators, respectively, as those given by equations (3.6)–(3.8). The summations are taken over all allowed states. If N is sufficiently large then $N/(N-1)$ may be replaced by 1.

3.2. The exchange interaction

It is a common fact that in a contemporary solid-state physics research an exchange interaction by far exceeds the role in a dynamical behavior of solids in comparison with the other types of interactions. Hence we devote the present section to the fundamental arguments leading to the foundation of this concept. Actually the exchange interaction is closely connected with the spin one-half, so we start by examining some general properties of the wave function in the presence of the spin variables. We know that the Schrödinger equation does not take the spin of the particles into account, but this by no means is the lack of a good description of the system. What is more, if the interactions among the particles do not depend upon the spin (e.g. an electrostatic interaction as long as a nonrelativistic approximation is used) then the hamiltonian of the system does not contain the spin variables. Therefore the total wave function may be written as a product of two wave functions, one containing the space coordinates, the other containing the spin variables,

$$\Phi(\boldsymbol{r}_1, \boldsymbol{S}_1, \ldots \boldsymbol{r}_N, \boldsymbol{S}_N) = \varkappa(\boldsymbol{r}_1, \ldots \boldsymbol{r}_N)\, \psi(\boldsymbol{S}_1, \ldots \boldsymbol{S}_N). \qquad (3.25)$$

It is clear that by using the Schrödinger equation we may determine only the coordinate wave function $\psi(\boldsymbol{r}_1, \ldots, \boldsymbol{r}_N)$, whereas the spin wave function $\chi(\boldsymbol{S}_1, \ldots, \boldsymbol{S}_N)$ remains arbitrary. Although the mutual interactions are taken to be spin independent the energy of the system may still depend on the total spin of the system. This is not in contradiction to the accepted view on the particles as identical objects.

Let us look at the system having two particles each of spin 1/2, such as two neighbor electrons moving in an atomic field of a ferromagnetic crystal, or two neighbor protons moving in a double-minimum potential field of a ferroelectric crystal. The following formulation is independent of the actual origin of the spin, it might be the real spin of an electron or the pseudo-spin of a ferroelectric proton. In any case the total wave function depends on the relative orientation of the individual spins. We denote the eigenvalue of the total spin by S and the eigenvalue of the z component by S_z. Therefore the total wave function satisfies the equations

$$\begin{aligned} \boldsymbol{S}^2\chi &= \hbar^2 S(S+1)\,\chi, \\ S_z\chi &= \hbar(S_{z1}+S_{z2})\chi, \end{aligned} \tag{3.26}$$

where S is equal to 0 or 1, and so we say that the electrons form a singlet or a triplet, respectively. Next we denote by α and β the spin wave functions which respectively correspond to the spin components $S_z = \frac{1}{2}$ and $-\frac{1}{2}$. Therefore the total wave functions are given by the following antisymmetric or symmetric combinations:

$$\chi_a(1, 2) = \frac{1}{\sqrt{2}}[\alpha(1)\beta(2)-\beta(1)\alpha(2)], \qquad S = S_z = 0, \tag{3.27}$$

$$\chi_s(1, 2) = \begin{cases} \alpha(1)\alpha(2), & (S, S_z) = (1, 1) \\ \dfrac{1}{\sqrt{2}}[\alpha(1)\beta(2)+\beta(1)\alpha(2)], & (S, S_z) = (1, 0) \\ \beta(1)\beta(2), & (S, S_z) = (1, -1). \end{cases} \tag{3.28}$$

Using the Pauli exclusion principle we may write the total wave function in the form which is in agreement with the Pauli theorem,[1-2]

$$\begin{aligned} \Phi_{12} &= \psi_s(\boldsymbol{r}_1, \boldsymbol{r}_2)\,\chi_a(1, 2), \quad \text{if} \quad S = 0, \\ \Phi_{12} &= \psi_a(\boldsymbol{r}_1, \boldsymbol{r}_2)\,\chi_s(1, 2), \quad \text{if} \quad S = 1, \end{aligned} \tag{3.29}$$

where

$$\Phi_{12} = \Phi(\boldsymbol{r}_1, \boldsymbol{r}_2, \boldsymbol{S}_1, \boldsymbol{S}_2).$$

Now we assume that a weak interaction of the form $V(\boldsymbol{r}_{12})$ acts among the particles where $\boldsymbol{r}_{12} = \boldsymbol{r}_2 - \boldsymbol{r}_1$. This assumption is valid so long as the particles are strongly bound to some major system, as the electrons are bound by diatomic molecules, or the protons are bound by a double-minimum potential field. In any case the motion of the particle must be localized to a tiny volume around the major system. Here we may use ordinary perturbation theory to calculate the mean energy as a first-order correction to the mean energy of the bound states. Hence, denoting the first-order correction term with $E_{(1)}$ we obtain

$$E_{(1)} = \sum_{\boldsymbol{S}_1, \boldsymbol{S}_2} \int d\tau_1 \int d\tau_2 \Phi_{12}^* V(\boldsymbol{r}_{12}) \Phi_{12} \tag{3.30}$$

where the integration is extended from $-\infty$ to ∞. Using the wave functions (3.29) we obtain two values for $E_{(1)}$ as follows:

$$E_{(1)} = \begin{cases} A+J, & \text{if} \quad S = 0, \\ A-J, & \text{if} \quad S = 1, \end{cases} \tag{3.31}$$

where the direct integral A and the exchange integral J are defined by

$$A = \int d\tau_1 \int d\tau_2 \, |\psi_1(\boldsymbol{r}_1)|^2 \, V(\boldsymbol{r}_{12}) \, |\psi_2(\boldsymbol{r}_2)|^2, \tag{3.32}$$

$$J = d\, d\tau_1 \int d\tau_2 \psi_1^*(r_1) \, \psi_2^*(r_2) \, V(r_{12}) \, \psi_2(\boldsymbol{r}_1) \, \psi_1(\boldsymbol{r}_2) \tag{3.33}$$

and the integration again is extended from $-\infty$ to ∞. We should emphasize that the exchange integral sometimes is called the "exchange" interaction and that it has no classical analogy thus entirely being of the quantum-mechanical origin.

The obtained results elucidate an interesting phenomenon. The energy levels corresponding to the symmetric solution of the Schrödinger equation may in fact be realized only when the spins of the electrons are "antiparallel" relative to one another, i.e. when $S = 0$, whereas the energy levels corresponding to the antisymmetric solution of the Schrödinger equation require the spins of the electrons be "parallel" to one another, i.e. $S = 1$. Clearly the energy levels depend on the magnitude of the total spin. From this viewpoint we may talk of a specific "exchange" interaction which would have disappeared if the spin of the electron disappeared. In this case the energy of the system would have been reduced to the direct integral A and the exchange integral J would have vanished and so would the difference between the symmetrical and antisymmetrical combinations. The exchange integral can be represented in the form of an operator. Indeed, we know that the matrix element of the operator $\boldsymbol{S}_1 \cdot \boldsymbol{S}_2$ depends on the magnitude S according to the expression

$$\boldsymbol{S}_1 \cdot \boldsymbol{S}_2 = \tfrac{1}{2}(\boldsymbol{S}^2 - \boldsymbol{S}_1^2 - \boldsymbol{S}_2^2) = \begin{cases} -\frac{3}{4} & \text{if} \quad S = 0, \\ \frac{1}{4} & \text{if} \quad S = 1, \end{cases} \tag{3.34}$$

so the exchange integral may be replaced by the operator

$$\mathcal{H}_{\text{exch}} = -\tfrac{1}{2} J(1 + 4\boldsymbol{S}_1 \cdot \boldsymbol{S}_2). \tag{3.35}$$

The reader may verify that the above operator has the matrix elements either J or $-J$, if S is equal to either 0 or 1, respectively.

Depending on the sign of J the exchange coupling is called either ferromagnetic if $J > 0$ or antiferromagnetic if $J < 0$ (see Table 3.1).

TABLE 3.1.

Matrix Elements and Spin Arrangement of an Exchange Interaction

Matrix element	Ground state	Excited state	Exchange integral	Spin arrangement
J	↑ ↑	↑ ↓	$J > 0$	Ferromagnetic
$-J$	↑ ↓	↑ ↑	$J < 0$	Antiferromagnetic

The question as to why the exchange integral is positive for one particular metal and negative for another is illuminated by Slater.[3] It appears that the condition for the occurrence of ferromagnetism or antiferromagnetism depends on the numerical ratio of the observed interatomic distance in a given crystal to the sum of radii for 3d electrons. Slater has estimated that the introduced ratio should be greater, but not very much greater, than 3 in order for the ferromagnetism to occur. On the other hand, if this ratio is smaller, but not very much smaller, than 3 then the antiferromagnetism can occur. These cases are illustrated in Table 3.2 and Table 3.3 for antiferromagnetic and ferromagnetic transition elements, respectively.

It is useful to paint a picture of the actual exchange process taking place in the course of time. Assuming that one electron at the moment $t = 0$ occupies the state ψ_1 (1) while the other electron at the same moment occupies the state ψ_2 (2), we may write the total wave function in the form

$$\Phi_{12}(t=0) = \frac{1}{\sqrt{2}}[\psi_s(1,2)+\psi_a(1,2)] = \psi_1(1)\,\psi_2(2). \qquad (3.36)$$

Here ψ_s and ψ_a correspond to the energy levels E_s and E_a, respectively where

$$E_{s,\,a} = E_0+A\pm J, \qquad (3.37)$$

E_0 being the unperturbed energy level, A and J being the expressions (3.32–3.33). Therefore the above wave functions have the following time dependence:

$$\psi_s(1, 2, t) = \psi_s(1, 2) \exp\{-i\hbar^{-1}(E_0+A+J)t\},$$
$$\psi_a(1, 2, t) = \psi_a(1, 2) \exp\{-i\hbar^{-1}(E_0+A-J)t\}. \qquad (3.38)$$

Hence the total wave function varies with time according to the law

$$\Phi_{12}(t) = \frac{1}{\sqrt{2}}[\psi_s(1, 2, t)+\psi_a(1, 2, t)]$$
$$= [\psi_1(1)\psi_2(2) \cos(\hbar^{-1} Jt)-i\psi_2(1)\,\psi_1(2) \sin(\hbar^{-1} Jt)]$$
$$\times\exp\{-i\hbar^{-1}(E_0+A)t\}. \qquad (3.39)$$

This shows that the two electrons will exchange their positions after some time interval τ_{exch}. If this interval extends to the moment when the electron 1 comes to the state ψ_2 while the electron 2 comes to the state ψ_1 then

$$\sin(\hbar^{-1} J\tau_{\text{exch}}) = 1$$

TABLE 3.2.

Conditions for the Occurrence of Antiferromagnetism‡

Bonding atoms	Description of structure	Bond distance, Å	Sum of radii of 3*d* electrons, Å	Ratio	Antiferromagnetic
Ti–Ti	metal (hex)	2.90–2.95	1.10	2.64–2.68	
	metal (bcc)	2.86		2.60	No
V–V	metal (bcc)	2.63	0.98	2.68	No
Cr–Cr	metal (bcc)	2.50	0.90	2.78	
	metal (hex)	2.71–2.72		3.01–3.02	Yes
Mn–Mn	metal (α)	2.24–3.00	0.84	2.67–3.57	
	metal (β)	2.36–2.67		2.81–3.18	
	metal (γ) (fcc)	2.67		3.18	Yes

‡ First four colums after Slater.[(3)]

TABLE 3.3.

Conditions for the Occurrence of Ferromagnetism‡

Bonding atoms	Description of structure	Bond distance, Å	Sum of radii of 3*d* electrons, Å	Ratio	Ferromagnetic?
Fe–Fe	metal (bcc)	2.48	0.78	3.18	
	metal (fcc)	2.54		3.26	Yes
Co–Co	metal (fcc)	2.51	0.72	3.49	
	metal (hex)	2.49–2.50		3.46–3.47	Yes
Ni–Ni	metal (fcc)	2.49	0.68	3.66	
	metal (hex)	2.65		3.90	Yes
Cu–Cu	metal (fcc)	2.54	0.64	3.97	No
Zn–Zn	metal (hex)	2.66–2.91	0.60	4.43–4.85	No

‡ Foirst fur columns after Slater.[3]

or

$$\tau_{\text{exch}} = \pi\hbar/(2J). \tag{3.40}$$

The square magnitude of the wave function

$$\Phi_{12}(t = \tau_{\text{exch}}) = i\psi_2(1)\,\psi_1(2)\exp\{-i\hbar^{-1}(E_0 + A)t\} \tag{3.41}$$

corresponds to the probability of finding the electron 1 at the position 2 whereas the electron 2 at the position 1. After the time $2\tau_{\text{exch}}$ has expired each electron comes back to its original position. Therefore we may say that τ_{exch} is just the time interval needed by the two neighbor electrons to exchange their equilibrium positions. The same conclusion is true when applied to the motion of the neighbor protons which surround a central phosphate group in a ferroelectric crystal. Here the proton is allowed to move around two equilibrium positions in a double-minimum potential field. If the equilibrium positions coincide with the minima of the potential field then the exchange process of a given pair of protons is exactly the same as that of a given pair of electrons in a ferromagnetic crystal. In both cases we must introduce an exchange interaction to account for the energy difference.

The above consideration makes clear the origin of the expression "pseudo-spin" used to describe both ferromagnetic and ferroelectric phenomena as far as the spontaneous arrangement is concerned at least.

A more exact approach to the problem of an exchange interaction in solids using the completely antisymmetrized wave function for the electron system is established by Van Vleck[4–5] and Herpin.[6]

3.3. The spin variables through the second quantization method

The spin variables in quantum mechanics are operators which obey the following commutation relations[7]

$$[S^2, S_z] = 0, \qquad [S_z, S_+] = \hbar S_+$$
$$[S_z, S] = -\hbar S_-, \qquad [S_+, S_-] = 2\hbar S_z \tag{3.42}$$

where

$$S_+ = S_x \pm iS_y. \tag{3.43}$$

We take the simultaneous set of normalized eigenvectors of the operators S^2 and S_z which mutually commute as the basic representation. We denote the corresponding eigenvalues with S and M whereas their common eigenvector we denote with $|SM\rangle$. Clearly, there are $2S+1$ different values for M,

$$-S \leqslant M \leqslant S$$

with $\Delta M = \pm 1$. Also we have for $\hbar = 1$

$$S^2 |SM\rangle = S(S+1)|SM\rangle,$$
$$S_z |SM\rangle = M|SM\rangle. \tag{3.44}$$

For a given value of S the eigenvectors $|SM\rangle$ span a $(2S+1)$-dimensional representation. These eigenvectors form a set of orthogonal unit vectors,

$$\langle SM|S'M'\rangle = \Delta(S-S')\,\Delta(M-M'). \tag{3.45}$$

It is the task of quantum mechanics to evaluate the nonvanishing matrix elements of the operators S_+ and S_-. Here we quote the final result:

$$S_+|SM\rangle = [(S-M)(S+M+1)]^{1/2}|SM+1\rangle$$
$$S_-|SM\rangle = [(S+M)(S-M+1)]^{1/2}|SM-1\rangle. \tag{3.46}$$

The simplest case is obtained by inserting $S = \frac{1}{2}$ in the above formulae. Taking $\hbar = 1$ we arrive at

$$\langle \tfrac{1}{2} M' | S_+ | \tfrac{1}{2} M \rangle = \begin{array}{c|cc} M' \backslash M & \tfrac{1}{2} & -\tfrac{1}{2} \\ \hline \tfrac{1}{2} & 0 & 1 \\ -\tfrac{1}{2} & 0 & 0, \end{array}$$

$$\langle \tfrac{1}{2} M' | S_- | \tfrac{1}{2} M \rangle = \begin{array}{c|cc} \tfrac{1}{2} & 0 & 0 \\ -\tfrac{1}{2} & 1 & 0, \end{array}$$

$$\langle \tfrac{1}{2} M' | S_z | \tfrac{1}{2} M \rangle = \begin{array}{c|cc} \tfrac{1}{2} & \tfrac{1}{2} & 0 \\ -\tfrac{1}{2} & 0 & -\tfrac{1}{2}. \end{array} \qquad (3.47)$$

The matrix elements on the right of equation (3.47) compose the matrix representation for S_+, S_-, and S_z, respectively.

Instead of dealing with the quantum number M we may introduce a quantum number n such that

$$n = S - M, \qquad \Delta n = -\Delta M. \qquad (3.48)$$

It is clear than n passes the set of values $n = 0, 1, \ldots, 2S$ when M passes the set of values $M = -S, \ldots, S$. The quantum number n gives therefore a departure of the z component S_z from the fixed value $S_z = S$. Using (3.48) the matrix elements for the spin components in units $\hbar = 1$ become

$$\begin{aligned} S_+ | S, n\rangle &= [(2S+1-n)\,n]^{1/2} | S, n-1\rangle, \\ S_- | S, n\rangle &= [(2S-n)(n+1)]^{1/2} | S, n+1\rangle, \\ S_z | S, n\rangle &= (S-n) | S, n\rangle. \end{aligned} \qquad (3.49)$$

The quantum number n may further be replaced by a set of creation and annihilation operators as follows:

$$\begin{aligned} n &= a^\dagger a, \\ a^\dagger | S, n\rangle &= \sqrt{(n+1)}\, | S, n+1\rangle, \\ a | S, n\rangle &= \sqrt{n}\, | S, n-1\rangle, \end{aligned} \qquad (3.50)$$

where the operator $a^\dagger$ "creates" a departure of the z component S_z from the fixed value $-S$, whereas the operator a "annihilates" a departure of this component from the fixed value $-S$. The minimum and maximum values for n determine the limiting eigenvectors $|S, n = 0\rangle$ and $|S, n = 2S\rangle$. These eigenvectors are characterized by

$$a|S, n = 0\rangle = a^\dagger |S, n = 2S\rangle = 0. \tag{3.51}$$

We shall call the state corresponding to the eigenvector $|S, n = 0\rangle$ the "vacuum" or "ground" state.

Using the introduced operators we may rewrite the matrix elements (3.49) in the form

$$\begin{aligned} S_+|S, n\rangle &= \sqrt{(2S)}\sqrt{(1-\varepsilon_+)}\,a|S, n\rangle, \\ S_-|S, n\rangle &= \sqrt{(2S)}\,a\sqrt{(1-\varepsilon_-)}|S, n\rangle, \\ S_z|S, n\rangle &= (S-a^\dagger a)|S, n\rangle \end{aligned} \tag{3.52}$$

where

$$\varepsilon_+ = \frac{n-1}{2S}, \qquad \varepsilon_- = \frac{n}{2S}. \tag{3.53}$$

It is easy to see by using the expressions (3.52–3.53) that the limiting values for the quantum number n are fully guaranted. Indeed, the minimum value $n = 0$ is guarranted by the existence of the vacuum state whereas the maximum value $n = 2S$ is guaranteed by the factor $\sqrt{(1-\varepsilon_-)}$ which vanishes at $n = 2S$. We emphasize here that the matrix elements (3.52–3.53) are expressed in a numerical form due to the fact that the factors $(1-\varepsilon_+)^{1/2}$ and $(1-\varepsilon_-)^{1/2}$ when applied to the eigenvectors $a|S, n\rangle$, or $|S, n\rangle$, act as the ordinary numbers.

However, we may present the matrix elements (3.49) in an alternative operational form. To do so we note that an eigenvalue of the operator $[1-a^\dagger a/(2S)]^{1/2}$ is equal to the square root of the eigenvalue of the operator $1-a^\dagger a/(2S)$. Therefore, the matrix elements (3.52–3.53) go over into

$$\begin{aligned} S_+|S, n\rangle &= \sqrt{(2S)}\sqrt{\{1-a^\dagger a/(2S)\}}\,a|S, n\rangle, \\ S_-|S, n\rangle &= \sqrt{(2S)}\,a^\dagger\sqrt{\{1-a^\dagger a/(2S)\}}|S, n\rangle, \\ S_z|S, n\rangle &= (S-a^\dagger a)|S, n\rangle. \end{aligned} \tag{3.54}$$

Also the commutation relation

$$S_+S_- - S_-S_+ = 2S_z$$

goes over into

$$2S\sqrt{\{1-a^\dagger a/(2S)\}}\, aa^\dagger \sqrt{\{1-a^\dagger a/(2S)\}} \\ -2S\, a^\dagger[1-a^\dagger a/(2S)]\, a = 2(S-a^\dagger a). \tag{3.55}$$

In an orthonormalized basis as built up of the eigenvectors of the occupation number operator $|n\rangle$, where $0 \leq n \leq 2S$, the transformation

$$[1-a^\dagger a/(2S)]^{1/2}$$

is diagonal having the eigenvalues

$$[1-n/(2S)]^{1/2}.$$

Therefore the commutation relation (3.55) leads to

$$aa^\dagger - a^\dagger a = 1,$$

or

$$[a, a^\dagger]_{(-)} = 1. \tag{3.56a}$$

Since the operators S_+ and S_- commute each with itself we find

$$[a, a]_{(-)} = [a^\dagger, a^\dagger]_{(-)} = 0. \tag{3.56b}$$

The present formalism can be applied to systems having a large number of spins such are the ferromagnetic or ferroelectric crystals. To do so we require the spin components at different lattice sites to commute one with another in all combinations,

$$[S_{\lambda j}, S_{\mu k}] = 0, \tag{3.57}$$

where λ, $\mu = +, -,$ or z, $j \neq k$ denote the lattice sites. Using the above formalism we obtain

$$[a_j, a_k^\dagger]_{(-)} = [a_j, a_k]_{(-)} = [a_j^\dagger, a_k^\dagger]_{(-)} = 0 \tag{3.58}$$

where $j \neq k$. The developed formalism may be applied to the exchange interaction of the type (3.35), or to any interaction involving the spin variables. Using the representation (3.54–3.58) such an interaction is expressed in terms of the creation and annihilation operators. For this reason we may look upon these operators as the amplitudes of some abstract field. The various terms entering the original hamiltonian will then give the single-particle energy as well as the mutual interaction among the particles in exactly the same manner by which a general interaction for a large system of particles was expressed by the second quantization method.

Using the developed formalism we actually deal with a gas consisting of some abstract quasi-particles whose statistical behavior is yet to be determined. A formal point of view suggests that such a gas consists of quasi-particles whose field amplitudes obey a set of commutation relations. Therefore it looks as if we were dealing with a system of bosons. However, such a boson system is not real owing to the fact that the occupation number operator $a_j^\dagger a_j$ is restricted to the eigenvalues $n_j = 0, 1, \ldots, 2S$, whereas the real boson system would require also the eigenvalues for $n_j > 2S$. As we know these latter values are unphysical posing a great obstacle to the developed formalism. However, there are two limiting situations where the painted boson picture presents a fair approximation to the real system, one being the case where S is large enough in order to justify the use of commutation relations for bosons, the other case being the dynamical behavior of the system in the ground state or very close to it. In both cases we may consider the operator $a^\dagger a/(2S)$ when applied to the eigenvectors $|S, n\rangle$ as a small quantity compared to unity. So we expand the square root in equations (3.54) in powers of $a^\dagger a/(2S)$ to obtain

$$
\begin{aligned}
S_{+j} &= \sqrt{(2S)}[1 - a_j^\dagger a_j/(4S)]\, a_j \ldots \\
S_{-j} &= \sqrt{(2S)}\, a_j^\dagger [1 - a_j^\dagger a_j/(4S)] + \ldots \\
S_{zj} &= S - a_j^\dagger a_j .
\end{aligned} \tag{3.59}
$$

The above expansion was used by Holstein and Primakoff[8] to investigate the thermodynamic properties of a Heisenberg exchange interaction acting between the ferromagnetic spins at low temperatures. In a current literature it is known as the Holstein–Primakoff representation. We may expect having used equations (3.59) to arrive at a fair approximation only if the thermal average $\langle a_j^\dagger a_j \rangle$ is much smaller than unity.

The present method is based on the assumption that a Hilbert space is associated with the system of ferromagnetic spins, the basic elements of the Hilbert space being identical to the physically pure states. Its general elements must therefore be identical to the physically mixed states. A thermal average of any operator depending on the spin variables is expressed in the Hilbert space by a certain scalar product.

The introduced formalism has a certain particle character in spite of having no resemblance to either boson or fermion picture. This specific character is particularly valid for the systems composed of the spin $\frac{1}{2}$. Indeed, using the matrix representation (3.47) we find

$$S_{zj} = \tfrac{1}{2} - S_{-j}S_{+j}$$

$$S_{+j}S_{-k} - S_{-k}S_{+j} = \begin{cases} 1 - 2S_{-j}S_{+j}, & \text{if} \quad j = k \\ 0, & \text{if} \quad j \neq k, \end{cases}$$

$$S_{-j}^2 = S_{+j}^2 = 0. \tag{3.60}$$

By a formal replacement

$$S_{-j} = p_j^\dagger, \quad S_{+j} = p_j. \tag{3.61}$$

we arrive at the combined commutation relation

$$p_j p_k^\dagger - p_k^\dagger p_j = (1 - 2p_j^\dagger p_j)\,\triangle(j-k),$$
$$p_j^2 = p_j^{\dagger 2} = 0. \tag{3.62}$$

The system of pseudo-spins obeying the commutation rules (3.60–3.62) is in the current literature[9] referred to as the system of "paulions". It is clear that the paulions behave like bosons at different lattice sites ($j \neq k$), whereas they behave like fermions at one and the same lattice site ($j = k$), i.e.

$$[p_j, p_k^\dagger]_{(-)} = 0, \qquad j \neq k,$$
$$[p_j, p_j^\dagger]_{(+)} = 1. \tag{3.63}$$

In the region of low temperatures the paulions behave like a gas of bosons, whereas we have no sufficient knowledge of their behavior in the region of high temperatures.

3.4. The Heisenberg ferromagnet in a boson representation

At low temperatures where the thermal average of the occupation number operator is small compared to 1 the ferromagnet with a Heisenberg exchange interaction acting between the ferromagnetic spins may be described by using a Holstein–Primakoff representation (3.59). We shall often refer to this set of assumptions as the "Heisenberg ferromagnet" at low temperatures. For the reason of clarity we repeat the basic assumptions as follows.

1. The interactions act between the nearest-neighbor ferromagnetic spins in a crystal lattice with a cubic symmetry and have the exchange type,

$$-\mathscr{H} = \sum J(\boldsymbol{R}-\boldsymbol{R}')\, \boldsymbol{S}_{\boldsymbol{R}} \cdot \boldsymbol{S}_{\boldsymbol{R}'} + g\mu_B H \sum S^z_{\boldsymbol{R}} \tag{3.64}$$

where $\boldsymbol{R}$ and $\boldsymbol{R}'$ denote the lattice sites, $\boldsymbol{S}_{\boldsymbol{R}}$ and $\boldsymbol{S}_{\boldsymbol{R}'}$ denote the spin variables of the magnitude S not necessarily restricted to $S = \frac{1}{2}$, μ_B is the magnitude of the Bohr magneton, g is an appropriate factor, H is a weak external magnetic field directed along the positive z axis. In what follows we explicitly consider only the first term in (3.64), thus having neglected the influence of the external magnetic field.

2. The thermal average $\langle a^\dagger_{\boldsymbol{R}} a_{\boldsymbol{R}} \rangle$ satisfies the condition

$$\langle a^\dagger_{\boldsymbol{R}} a_{\boldsymbol{R}} \rangle \ll 1 \tag{3.65}$$

and the spin variables are taken as follows:

$$\begin{aligned} S^+_{\boldsymbol{R}} &\cong \sqrt{(2S)}[1 - a^\dagger_{\boldsymbol{R}} a_{\boldsymbol{R}}/(4S)]\, a_{\boldsymbol{R}}, \\ S^-_{\boldsymbol{R}} &\cong \sqrt{(2S)}\, a^\dagger_{\boldsymbol{R}}[1 - a^\dagger_{\boldsymbol{R}} a_{\boldsymbol{R}}/(4S)], \\ S^z_{\boldsymbol{R}} &= S - a^\dagger_{\boldsymbol{R}} a_{\boldsymbol{R}}. \end{aligned} \tag{3.66}$$

If the approximation (3.65) is accepted then the commutation relation

$$[S_R^+, S_R] = 2S_R^z \Delta(R-R')$$

leads to

$$[a_R, a_{R'}^\dagger]_{(-)} \simeq \Delta(R-R'). \tag{3.67}$$

This shows that the system of the ferromagnetic spins in the ground state, or in states close to the ground state, is reduced to the system of quasi-particles having a boson character. Our next task is to formulate the elementary excitations for such a boson system and then to evaluate, using the energy of the elementary excitations, the spontaneous magnetization and specific heat.

To do this we introduce the Fourier transform for the creation and annihilation operators

$$a_R^\dagger = N^{-1/2} \sum_q \exp(i\boldsymbol{q}\cdot\boldsymbol{R})\, a_q^\dagger,$$

$$a_R = N^{-1/2} \sum_q \exp(-i\boldsymbol{q}\cdot\boldsymbol{R})\, a_q, \tag{3.68}$$

where $\boldsymbol{q}$ denotes a reciprocal lattice vector, N is the total number of ferromagnetic spins. In the transformed hamiltonian (3.64) there will appear the expressions like

$$J(\boldsymbol{q}) = \sum J(\boldsymbol{R}-\boldsymbol{R}') \exp[i\boldsymbol{q}(\boldsymbol{R}-\boldsymbol{R}')] = J(\boldsymbol{q})^* = J(-\boldsymbol{q}). \tag{3.69}$$

Such expressions are already calculated in Appendix A for the crystal lattices having a cubic symmetry and for the nearest-neighbor interactions. As $\boldsymbol{q} \to 0$ the expansion is given by

$$J(\boldsymbol{q}) = J\{1-(qa)^2/z + A(\phi, \theta)(qa)^4 - B(\phi, \theta)(qa)^6 \ldots\} \tag{3.70}$$

where z denotes the number of nearest neighbors, a is the lattice constant, whereas J denotes an effective exchange integral.

The exchange operator $\boldsymbol{S}_R \cdot \boldsymbol{S}_{R'}$ may be written

$$\boldsymbol{S}_R \cdot \boldsymbol{S}_{R'} = \tfrac{1}{2}(S_R^- S_{R'}^+ + S_R^+ S_{R'}^-) + S_R^z S_{R'}^z, \tag{3.71}$$

so by having neglected the sixth and higher-order terms we obtain

$$
\begin{aligned}
S_R^- S_{R'}^+ &= 2S[a_R^\dagger a_{R'} - a_R^\dagger a_{R'}^\dagger a_{R'} a_{R'}/(4S) \\
&\quad - a_R^\dagger a_R^\dagger a_R a_{R'}/(4S)] + \dots, \\
S_R^+ S_{R'}^- &= 2S[a_R a_{R'}^\dagger - a_R^\dagger{}_{\to} a_R{}_{\to}{}_R{}_{\to} a_{R'}^\dagger/(4S) \\
&\quad - a_R{}_{\to} a_{R'}^\dagger a_{R'}^\dagger{}_{\to} a_{R'}/(4S)] + \dots, \\
S_R S_{R'}^z &= S^2 - S(a_R^\dagger a_R + a_{R'}^\dagger a_{R'}) + a_R^\dagger a_R a_{R'}^\dagger a_{R'}.
\end{aligned} \tag{3.72}
$$

The hamiltonian (3.64) is transformed into

$$
\begin{aligned}
\mathcal{H} = &-S^2 \sum J(\boldsymbol{R}-\boldsymbol{R}') + 2S \sum J(\boldsymbol{R}-\boldsymbol{R}')\,(a_R^\dagger a_R - a_R^\dagger a_{R'}) \\
&- J(\boldsymbol{R}-\boldsymbol{R}')\, a_R^\dagger a_R a_{R'}^\dagger a_{R'} + \tfrac{1}{2} \sum J(\boldsymbol{R}-\boldsymbol{R}')\,(a_R^\dagger a_{R'}^\dagger a_{R'} a_{R'} \\
&+ a_R^\dagger a_R^\dagger a_R a_{R'}).
\end{aligned} \tag{3.73}
$$

We can now transform every term appearing in the above hamiltonian with the help of the Fourier transform (3.68). Having in mind that each interaction pair is counted only once we obtain the following result:

$$
\sum J(\boldsymbol{R}-\boldsymbol{R}') = zNJ, \tag{3.74a}
$$

$$
\begin{aligned}
\sum J(\boldsymbol{R}-\boldsymbol{R}')\, a_R^\dagger a_{R'} &= N^{-1} \sum J(\boldsymbol{R}-\boldsymbol{R}')\, a_q^\dagger a_{q'} \exp[i(\boldsymbol{q}\cdot\boldsymbol{R} - \boldsymbol{q}'\cdot\boldsymbol{R}')] \\
&= N^{-1} \sum zJ(\boldsymbol{q})\, a_q^\dagger a_{q+p} \exp(-i\boldsymbol{p}\cdot\boldsymbol{R}') \\
&= \sum zJ(\boldsymbol{q})\, a_q^\dagger a_q,
\end{aligned} \tag{3.74b}
$$

where we have used the substitution $\boldsymbol{q}' = \boldsymbol{q}+\boldsymbol{p}$ and

$$
\sum_{R'} \exp(-i\boldsymbol{P}\cdot\boldsymbol{R}') = N\,\Delta(\boldsymbol{p}) = \begin{cases} N & \text{if} \quad \boldsymbol{p} = 0, \\ 0, & \text{if} \quad \boldsymbol{p} \neq 0, \end{cases} \tag{3.74c}
$$

which follows from the translational property for the reciprocal lattice vectors. Also the following relationship holds:

$$
\begin{aligned}
\sum J(\boldsymbol{R}-\boldsymbol{R}')\, a_R^\dagger a_R &= N^{-1} \sum J(\boldsymbol{R}-\boldsymbol{R}')\, a_q^\dagger a_{q'} \exp[i(\boldsymbol{q}-\boldsymbol{q}')\cdot\boldsymbol{R}] \\
&= \sum zJ a_q^\dagger a_q.
\end{aligned} \tag{3.74d}
$$

Next we transform the remaining three terms in (3.73) which depend on four operators. The first such term is equal to

$$\sum_{\text{all } \boldsymbol{R}} J(\boldsymbol{R}-\boldsymbol{R}')\, a_{\boldsymbol{R}}^{\dagger} a_{\boldsymbol{R}} a_{\boldsymbol{R}'}^{\dagger} a_{\boldsymbol{R}'} = N^{-2} \sum_{\text{all } \boldsymbol{R}, \text{ all } \boldsymbol{q}} J(\boldsymbol{R}-\boldsymbol{R}')\, a_{\boldsymbol{q}}^{\dagger} a_{\boldsymbol{q}'} a_{\boldsymbol{q}''}^{\dagger} a_{\boldsymbol{q}'''} \times$$
$$\times \exp\,[i(\boldsymbol{q}\cdot\boldsymbol{R}-\boldsymbol{q}'\cdot\boldsymbol{R}+\boldsymbol{q}''\cdot\boldsymbol{R}'-\boldsymbol{q}'''\cdot\boldsymbol{R}')].$$

Here we introduce the substitution

$$\boldsymbol{q}'' = \boldsymbol{q}'+\boldsymbol{p}, \qquad \boldsymbol{q}''' = \boldsymbol{q}+\boldsymbol{p},$$

which leads the above exponential factor to

$$\exp\,[i(\boldsymbol{q}-\boldsymbol{q}')\,(\boldsymbol{R}-\boldsymbol{R}')]\times\exp\,[i(\boldsymbol{p}-\boldsymbol{p}')\cdot\boldsymbol{R}'],$$

and the above term becomes

$$= N^{-2} \sum_{\boldsymbol{R}'\boldsymbol{q},\, \boldsymbol{q}',\, \boldsymbol{p},\, \boldsymbol{p}'} J(\boldsymbol{q}-\boldsymbol{q}')\, a_{\boldsymbol{q}}^{\dagger} a_{\boldsymbol{q}'} a_{\boldsymbol{q}'+\boldsymbol{p}'}^{\dagger} a_{\boldsymbol{q}+\boldsymbol{p}}$$
$$\times \exp\,[i(\boldsymbol{p}-\boldsymbol{p}')\cdot\boldsymbol{R}'].$$

The sum over $\boldsymbol{R}'$ may be performed immediately giving the factor $N\,\Delta(\boldsymbol{p}-\boldsymbol{p}')$; therefore, the above term becomes

$$= N^{-1} \sum_{\boldsymbol{q},\, \boldsymbol{q}',\, \boldsymbol{p}} zJ(\boldsymbol{q}-\boldsymbol{q}')\, a_{\boldsymbol{q}}^{\dagger} a_{\boldsymbol{q}'} a_{\boldsymbol{q}'+\boldsymbol{p}'}^{\dagger} a_{\boldsymbol{q}+\boldsymbol{p}}.$$

There are two important contributions coming from the above term, one with $\boldsymbol{p} = 0$, the other with $\boldsymbol{q} = \boldsymbol{q}'$. The former contribution is equal to

$$= N^{-1} \sum_{\boldsymbol{q},\, \boldsymbol{q}'} zJ(\boldsymbol{q}-\boldsymbol{q}')\, a_{\boldsymbol{q}}^{\dagger} a_{\boldsymbol{q}'}^{\dagger} a_{\boldsymbol{q}} a_{\boldsymbol{q}'} + N^{-1} \sum_{\boldsymbol{q},\, \boldsymbol{q}'} zJ(\boldsymbol{q}-\boldsymbol{q}')\, a_{\boldsymbol{q}}^{\dagger} a_{\boldsymbol{q}}. \quad (3.74\text{e})$$

The latter contribution is equal to

$$= N^{-1} \sum_{\boldsymbol{q},\, \boldsymbol{p}} zJ a_{\boldsymbol{q}}^{\dagger} a_{\boldsymbol{q}+\boldsymbol{p}}^{\dagger} a_{\boldsymbol{q}} a_{\boldsymbol{q}+\boldsymbol{p}} + N^{-1} \sum_{\boldsymbol{q},\, \boldsymbol{p}} zJ a_{\boldsymbol{q}}^{\dagger} a_{\boldsymbol{q}+\boldsymbol{p}}\, \Delta(\boldsymbol{p}). \quad (3.74\text{f})$$

By having neglected the second sum in equations (3.74a–f) we finally obtain

$$\sum_{\boldsymbol{R},\, \boldsymbol{R}'} J(\boldsymbol{R}-\boldsymbol{R}')\, a_{\boldsymbol{R}}^{\dagger} a_{\boldsymbol{R}} a_{\boldsymbol{R}'}^{\dagger} a_{\boldsymbol{R}'} = N^{-1} \sum_{\boldsymbol{q},\, \boldsymbol{q}'} z[J+J(\boldsymbol{q}-\boldsymbol{q}')]\times$$
$$\times a_{\boldsymbol{q}}^{\dagger} a_{\boldsymbol{q}'}^{\dagger} a_{\boldsymbol{q}} a_{\boldsymbol{q}'}. \quad (3.75)$$

Using a similar set of transformations we arrive at the result

$$\sum_{R, R'} J(R-R')\, a^{\dagger}_{R} a^{\dagger}_{R'} a_{R'} a_{R'} = \sum_{R, R'} J(R-R')\, a^{\dagger}_{R} a^{\dagger}_{R'} a_{R} a_{R'}$$
$$= N^{-1} \sum_{q, q'} z[J(q)+J(q')]\, a^{\dagger}_{q} a^{\dagger}_{q'} a_{q} a_{q'}. \tag{3.76}$$

Now we can write the hamiltonian (3.73) in the transformed form as follows:

$$\mathcal{H} = E_0 + 2S \sum_{q} z[J-J(q)]\, a^{\dagger}_{q} a_{q}$$
$$-N^{-1} \sum_{q, q'} z[J+J(q-q')-J(q)-J(q')]\, a^{\dagger}_{q} a^{\dagger}_{q'} a_{q} a_{q'}. \tag{3.77}$$

Here E_0 stands for the energy of the ground state,

$$E_0 = -zNS^2J. \tag{3.78}$$

Writing the above hamiltonian as

$$\mathcal{H} - E_0 = \sum_{q} E(q)\, a^{\dagger}_{q} a_{q} - N^{-1} \sum_{q, q'} V(q, q')\, a^{\dagger}_{q} a^{\dagger}_{q'} a_{q} a_{q'} \tag{3.79}$$

where

$$E(q) = 2zS[J-J(q)], \tag{3.80}$$
$$V(q, q') = z[(J+J(q-q')-J(q)-J(q')], \tag{3.81}$$

we can immediately see the physical meaning of the terms involved.

1. The first term gives the energy of the elementary excitations at zero temperature exactly, so $E = E(q)$ is a dispersion law. These elementary excitations are called "magnons", and the energy of a single magnon is just $E(q)$. At low temperatures the system of magnons behaves like a system of bosons. Indeed, using equations (3.67–3.68) we obtain in the low-temperature limit

$$[a_q, a^{\dagger}_{q'}]_{(-)} \simeq \Delta(q-q'), \tag{3.82}$$

which shows that the magnon gas in a thermodynamic sense looks like a boson gas. In the limiting case where $q \to 0$ the dispersion law for magnons is equal to‡

$$E(q) \simeq 2SJa^2q^2, \tag{3.83}$$

‡ Note that many authors in a current literature often express the dispersion law using the nearest-neighbor distance b instead of the lattice constant a. It is clear that the following relation holds for cubic lattices: $6a^2 = zb^2$.

where a denotes the lattice constant. Had we included an external magnetic field the above dispersion law would become

$$E(\boldsymbol{q}) \simeq g\mu_B H + 2SJa^2q^2 \tag{3.84}$$

where the additional term $g\mu_B H$ is sometimes referred to as the Zeeman splitting energy.

2. The second term in the hamiltonian (3.79) describes a dynamical interaction which is often named the "magnon–magnon" interaction. Expanded in a power series around the point $\boldsymbol{q} = \boldsymbol{q}' = 0$ for a cubic lattice this interaction is expressed in the form of a polynomial in $\boldsymbol{q}$ and $\boldsymbol{q}'$. Using the expansions

$$\begin{aligned} J(\boldsymbol{q}) &\simeq J\{1-(qa)^2/z + A(\phi, \theta)\,(qa)^4 - \ldots\} \\ J(\boldsymbol{q}') &\simeq J\{1-(q'a)^2/z + A(\phi', \theta')\,(q'a)^4 - \ldots\} \end{aligned} \tag{3.85}$$

where ϕ, θ, ϕ' and θ' denote the angles closed by the vectors $\boldsymbol{q}$ and $\boldsymbol{q}'$ with the system of cartesian coordinates, respectively, we may write the magnon interaction as follows:

$$\begin{aligned} V(\boldsymbol{q}, \boldsymbol{q}') \simeq J\{&2qq'a^2 \cos\gamma + zA(\alpha, \beta)\,|\boldsymbol{q}-\boldsymbol{q}'|^4 a^4 \\ &- zA(\phi, \theta)\,(qa)^4 - zA(\phi', \theta')\,(q'a)^4\} \end{aligned} \tag{3.86}$$

where α, β denote the angles closed by the vector $(\boldsymbol{q}-\boldsymbol{q}')$ with the system of cartesian coordinates, whereas γ is the angle closed by the vectors $\boldsymbol{q}$ and $\boldsymbol{q}'$ themselves. Very often we may use an average over the angle γ in performing a kind of numerical analysis. Such an average value leads to

$$\langle V(\boldsymbol{q}, \boldsymbol{q}')\rangle = \frac{1}{2\pi}\int_0^{2\pi} V(\boldsymbol{q}, \boldsymbol{q}')\,d\gamma. \tag{3.87}$$

Using the approximation (3.86) we arrive at the result

$$\begin{aligned} \langle V(\boldsymbol{q}, \boldsymbol{q}')\rangle \simeq zJ\{&A(\alpha, \beta)\,(q^4+q'^4+4q^2q'^2)\,a^4 \\ &- A(\phi, \theta)\,(qa)^4 - A(\phi', \theta')\,(q'a)^4\}. \end{aligned} \tag{3.88}$$

Here we have used the integrals

$$\int_0^{2\pi} (\cos\gamma)^{2n+1}\, d\gamma = 0,$$

$$\int_0^{2\pi} (\cos\gamma)^{2n}\, d\gamma = \begin{cases} \pi, & n = 1 \\ \dfrac{(2n-1)(2n-3)\ldots 3}{2n(2n-2)\ldots 4}\pi, & n \geqslant 2. \end{cases} \tag{3.89}$$

3.5. The spontaneous magnetization and specific heat

In the present section we shall apply the developed methods to the Heisenberg exchange interaction in order to evaluate the exact dispersion law, spontaneous magnetization and specific heat at low temperatures. In place of an approximate expression (3.83) which is valid only at $T = 0$ and for $\boldsymbol{q} \to 0$ we define the general dispersion law in the form

$$E(T, \boldsymbol{q}) = 2SJ \sum_{n=1}^{\infty} C_{(2n)}(1-\eta_{(2n)})\,(qa)^{2n} \tag{3.90}$$

where $C_{(2n)}$ are certain expansion coefficients, and $\eta_{(2n)}$ is a function of temperature,

$$\eta_{(2n)} = \eta_{(2n)}(T). \tag{3.91}$$

If all $\eta_{(2n)} = 0$ then the dispersion law (3.90) agrees with the expansion (3.80).

Now the spontaneous magnetization is defined by a thermal average of the z component of the magnetic moment and summed over a unit volume of the crystal,

$$M(T) = M(0)\langle S_{\boldsymbol{R}}^z\rangle/S \tag{3.92a}$$

or

$$M(T) = M(0)\,[1-\langle a_{\boldsymbol{R}}^{\dagger}a_{\boldsymbol{R}}\rangle/S]. \tag{3.92b}$$

Here equation (3.66) is used to describe the departure of the thermal average $\langle S^z_{\boldsymbol{R}} \rangle$ from an ideal value. Using the Fourier transform we arrive at

$$\langle a^\dagger_{\boldsymbol{R}} a_{\boldsymbol{R}} \rangle = N^{-1} \sum_{\boldsymbol{q}, \boldsymbol{q}'} \exp\,[i(\boldsymbol{q}-\boldsymbol{q}')\boldsymbol{\cdot}\boldsymbol{R}]\,\langle a^\dagger_{\boldsymbol{q}} a_{\boldsymbol{q}'} \rangle. \tag{3.93}$$

Clearly the matrix elements involved in equation (3.93) are those with $\boldsymbol{q} = \boldsymbol{q}'$, hence we may write

$$\langle a^\dagger_{\boldsymbol{q}} a_{\boldsymbol{q}'} \rangle = \frac{\Delta(\boldsymbol{q}-\boldsymbol{q}')}{\exp\,[\beta E(T, \boldsymbol{q})]-1}, \tag{3.94}$$

where the dispersion law appearing in the denominator of equation (3.94) is given by the expansion (3.90). By a self-consistent procedure any equation leading to an analytic expression for $E(T, \boldsymbol{q})$ must be given through the use of the hamiltonian (3.79). Therefore writing the dispersion law in the form

$$E(T, q) = 2SJ \sum_{n=1}^{\infty} C_{(2n)}(qa)^{2n} - N^{-1} \sum_{q, q'} V(q, q')\,\langle a^\dagger_{q} a_{q'} \rangle \tag{3.95}$$

and making a comparison to the original expression (3.90) we obtain a self-consistent method for the dispersion law and spontaneous magnetization in a closed and integral form. The written equations so far are correct at any temperature provided that a magnon–magnon interaction entering the hamiltonian (3.79) is sufficient to account for a complete dynamical behavior of the system of ferromagnetic spins which act through an exchange interaction.

Next we use an approximate expression for the expansion parameter $\eta_{(2n)}(T)$,

$$\eta_{(2n)} \equiv \eta, \qquad \eta_{(2n)} = 0, \qquad \text{if} \qquad n \geqslant 2. \tag{3.96}$$

The lowest-order approximation leads to

$$E(T, \boldsymbol{q}) \simeq 2SJ[(1-\eta)\,(qa)^2 - zA(\phi, \theta)\,(qa)^4 + zB(\phi, \theta)\,(qa)^6 + \ldots] \tag{3.97}$$

where

$$\eta = \frac{Va^2}{12\pi^2 NS} \int_0^\infty \frac{q^4 dq}{\exp[(1-\eta)\, a^2 q^2/\tau] - 1} \tag{3.98a}$$

τ is a dimensionless temperature defined by

$$\tau = kT/(2SJ). \tag{3.98b}$$

In the region of low temperatures the result of integrations is given with the help of Appendix C

$$\eta = \frac{V}{24\pi^2 N a^3 S} (1-\eta)^{-5/2} \tau^{5/2} \Gamma(\tfrac{5}{2})\, \zeta(\tfrac{5}{2}) \tag{3.98c}$$

where V denotes the volume of the crystal, Γ is a gamma function, and ζ is the Riemann zeta function. We can observe that a low-temperature expansion for the parameter η gives

$$\eta \propto T^{5/2}.$$

The spontaneous magnetization is determined by a low-temperature expansion. By equations (3.93–3.98) we obtain‡

$$\langle a_R^\dagger a_R \rangle = \frac{V}{(2\pi)^3 N} \int d\Omega \int_0^\infty \frac{q^2 dq}{\exp[(1-\eta) a^2 q^2/\tau] - 1} \tag{3.99a}$$

where $d\Omega = \sin\theta d\theta d\phi$. Having performed the indicated integrations we arrive at

$$\langle a_R^\dagger a_R \rangle = \frac{V}{4\pi^2 N} (1-\eta)^{-3/2}\, \Gamma(\tfrac{3}{2})\, \zeta(\tfrac{3}{2})\, \tau^{3/2}. \tag{3.99b}$$

‡ Here according to Bogoliubov and Shirkov[10] we make a transition from a discrete momentum representation to the continuous momentum representation by using the prescription

$$N^{-1} \sum_q \langle n_q \rangle = N^{-1} (L/2\pi)^3 \int 4\pi \langle n_q \rangle\, q^2\, dq = \frac{V}{2N\pi^2} \int \langle n_q \rangle\, q^2\, dq$$

where $V = L^3$ is the volume of the crystal.

Clearly the lowest-order expansion term is proportional to $T^{3/2}$. Had we taken the neglected terms of the order q^4, or even q^6, from the expansion (3.97) we would have arrived at the corrections to $\langle a_R^\dagger a_R \rangle$ proportional to $T^{5/2}$, or even $T^{7/2}$, respectively. Apart from the mentioned contributions the most interesting one comes from the expansion (3.99b) assuming that η is a small parameter. It is easy to see that such an expansion would lead to the term proportional to T^4 in a low-temperature region which would then represent a correction coming from the dynamical magnon interactions.

So the magnetization at low temperatures has a general behavior

$$M(T) = M(0)\{1-[a_0\tau^{3/2}+a_1\tau^{5/2}+a_2\tau^{7/2}\ a_3\tau^4+(\tau^{9/2})]\} \qquad (3.100)$$

where the coefficients a_j are all positive having a descending order of magnitude; here τ is a dimensionless temperature defined by (3.98b). Having performed the indicated integration one arrives at

$$a_j = \frac{V}{(2\pi)^3 NS} b_j \qquad (3.101)$$

with b_j being the expansion coefficients from Appendix C. Numerical analysis based on the present model is given in the following tables; in particular, Table 3.4 contains the transition (Curie) ferromagnetic

TABLE 3.4.

Transition Temperature of Ferromagnetic Elements

(after Kittel[11])

	Co	Dy	Fe	Gd	Ni
T_c, K	1400	85	1043	292	631

temperature for a number of metals, whereas the expansion coefficients a_j for cubic lattices and the saturation magnetization for Fe, Co, and Ni are given in Table 3.5 and Table 3.6, respectively. The a_0 term appearing in equation (3.100) represents a result identical to that of

TABLE 3.5.

Expansion Coefficients a_j for Cubic Lattices

	First factor			Second factor
	P	I	F	
a_0	$\frac{1}{8S}$	$\frac{1}{16S}$	$\frac{1}{32S}$	$\pi^{-3/2}\zeta(\frac{3}{2})$
a_1	$\frac{3}{128S}$	$\frac{9}{1024S}$	$\frac{3}{1024S}$	$\pi^{-3/2}\zeta(\frac{5}{2})$
a_2	$\frac{265}{504\times 64S}$	$\frac{2257}{1024\times 1008S}$	$\frac{265}{2016\times 256S}$	$\pi^{-3/2}\zeta(\frac{7}{2})$
a_3	$\frac{3}{512S^2}$	$\frac{3}{1024S^2}$	$\frac{3}{2048S^2}$	$\pi^{-3}\zeta(\frac{3}{2})\zeta(\frac{5}{2})$

Bloch,[15] whereas higher-order terms agree with those of Dyson,[16] the latter results being expressed in terms of a dimensionless temperature θ for which the present dimensionless temperature τ is related by

$$\tau = 4\pi\theta.$$

The specific heat is given by

$$C(T) = \frac{\partial\langle\mathcal{H}\rangle}{\partial T} \tag{3.102}$$

where

$$\langle\mathcal{H}\rangle = N^{-1}\sum_{\boldsymbol{q}} E(T, \boldsymbol{q})\,\langle n_{\boldsymbol{q}}\rangle. \tag{3.103}$$

It is a fair low-temperature approximation to take

$$E(T, \boldsymbol{q}) = 2SJ(1-\alpha\tau^{5/2})\,(qa)^2 \tag{3.104}$$

for all cubic lattices with

$$\alpha = \frac{\zeta(\frac{5}{2})}{32S\pi^{3/2}}, \tag{3.105}$$

a being the lattice constant. Using the above expansion we arrive at

TABLE 3.6.

Saturation Magnetization of Several Ferromagnetic Elements

	Observed at 0 K			Observed at 20 °C			Lattice structure‖	τ¶‖	Calculated $\frac{\sigma_0-\sigma_s}{\sigma_0}$
	σ_0‡	n_B‡	S§	σ_s‡	M_s‡	$\frac{\sigma_0-\sigma_s}{\sigma_0}$			
Fe	221.9	2.219	1	218.0	1714	1.76×10^{-2}	Cubic I	$16T/3T_c$	6.30×10^{-2}
							Cubic F	$8T/T_c$	6.07×10^{-2}
Co	162.5	1.715	1	161	1422	0.92×10^{-2}	Cubic F	$8T/T_c$	3.62×10^{-2}
			1/2				Cubic F	$6T/T_c$	4.58×10^{-2}
Ni	57.5	0.604	1/2	54.39	484.1	5.41×10^{-2}	Cubic F	$6T/T_c$	20.18×10^{-2}

‡ After Bozorth.(12) § After Lomer and Marshal.(13) ‖ After Slater.(14)

¶‖ By definition $\tau = \frac{z(S+1)\,T}{3T_c}$.

the result

$$\langle \mathcal{H} \rangle = \frac{3V\,\zeta(\frac{5}{2})}{16Na^3\pi^{3/2}}\,kT\left(\frac{\tau}{1-\alpha\tau^{5/2}}\right)^{3/2},$$

$$C(T) = c\tau^{3/2}+d\tau^4, \tag{3.106}$$

where

$$c = \frac{15V\,\zeta(\frac{5}{2})}{32Na^3\pi^{3/2}}\,k,$$

$$d = 3c\alpha. \tag{3.107}$$

It is worth noting that the specific heat in the present approximation is expressed by two contributions, one coming from the energy of free spin waves and the other coming from a spin-wave dynamical interaction. These terms are respectively proportional to $\tau^{3/2}$ and τ^4.

3.6. A review of genuine papers devoted to spin waves

The problem of obtaining a correct boson representation for the spin operators is an old one and it has the origin in the field of magnetism. In the theory of spin waves there exists a problem which has attracted so much attention and has stimulated a discussion in order to investigate the very basis of the theory. To the best of our knowledge the magnetization at low temperatures contains terms of the form $T^{3/2}$, $T^{5/2}$, and $T^{7/2}$ which correspond to the terms q^2, q^4, and q^6 of the dispersion law, respectively. Now the question may be formulated as follows: Apart from contributions having the form $T^{5/2}$, or $T^{7/2}$, what is the next degree of T whose contribution to the magnetization at low temperatures in a ferromagnet with a Heisenberg exchange interaction comes entirely from a nonlinear theory? As we know according to Holstein and Primakoff the problem is reduced to that of replacing the nonlinear operators S_+, S_-, and S_z with a set of approximate expansion equations of the form (3.59). It is understood in the framework of statistical mechanics that a thermal average $\langle a^\dagger a/2S\rangle$ is considerably smaller than 1 making the above expansions justified in a practical application to low temperatures. However, the elementary excitations at temperatures $T \sim T_c$ are far from the ground state, so there will appear unphysical states ($a^\dagger a > 2S$) giving rise to a number of principal difficulties.

Van Kranendonk[17] was one of the first researchers who tackled the present problem in a rigorous way. In two exhaustive papers he treated the spin deviations as a gas of interacting bosons to obtain the low-temperature expansion for the magnetization in terms of $T^{3/2}$, $T^{5/2}$, and $T^{7/2}$ with the coefficients for which one can show to be identical with those of Dyson but with a dynamical contribution of the form $T^{7/4}$ rather than T^4. The whole subject of spin waves in a Heisenberg ferromagnet was thoroughly investigated by Dyson[16] who developed a method of linked clusters using a power expansion for the partition function. By separating the kinematical terms coming from the use of a nonorthogonal set of spin waves from the dynamical terms he was able to show that this method leads to either powers of T or terms where the factor $\exp(-J/kT)$ dominates. All such terms vanish at low temperatures so that the leading contribution has the form T^4. As later demonstrated by Maleev,[18] Dyson in his analysis has actually avoided the use of the square operator $\sqrt{\{1-a^\dagger a/(2S)\}}$, but instead has used the following expansion:

$$\begin{aligned} S_+ &= \sqrt{(2S)}\, a, \\ S_- &= \sqrt{(2S)}[1-a^\dagger a/(2S)]a, \\ S_z &= a^\dagger a - S. \end{aligned} \tag{3.108}$$

In the contemporary literature the above transformation is often referred to as the Dyson–Maleev representation. As it can be observed the above transformation does not contain the famous square root from the operator $1-a^\dagger a/(2S)$; however, the spin variables S_+ and S_- are no longer hermitian conjugate to each other. It should be emphasized that the representation (3.108) satisfies exactly the commutation relations for the spin components.

After Dyson's work has appeared where a leading dynamical correction was shown to have the form T^4, many authors arrived at essentially the same result by using various methods. It is very instructive to mention some of them since every one is original and interesting.

One of the mentioned methods is due to Morita[19] who starts from the Holstein–Primakoff representation for the spin operators and then takes the following expansion

$$[1-a^\dagger a/(2S)]^{1/2} = \sum_{n=0}^{\infty} c_n a^{\dagger n} a^n/n!, \tag{3.109}$$

where the expansion coefficients c_n must be determined from the condition

$$f(a^\dagger a) = f(n) = \sum_{r=0}^{\infty} \frac{(-1)^r}{r!} a^{\dagger r} a^r \, [\Delta_r f(n)]_{n=0} \tag{3.110}$$

Here Δ_r denotes an rth derivation with respect to n. A fraction of the above summation up to $n = 2S$ will have the same effect as the square root, whereas for $n > 2S$ it becomes a complex number. Morita in fact takes into account terms up to $n = 2S$ and then adds to the original hamiltonian a spin potential of the form

$$v_0 \sum_{i=1}^{N} a_i^{\dagger(2S+1)} a_i^{(2S+1)}$$

where v_0 at the end of the calculation is allowed to take an infinite value. This term in effect prevents the system occupying the unphysical states with $n_i > 2S+1$. Using this method Morita has colculated the partition function where more and more pairs of interacting spin waves are explicitly considered to arrive at the dynamical correction T^4 with a high degree of accuracy. The coefficient in front of T^4 was not given numerically, so it is hard to conclude whether Morita's result is identical to that of Dyson.

Oguchi[20] has developed a relatively simple method which gives the correction terms for both the ferromagnet and antiferromagnet, the coefficient in a T^4 term for the ferromagnet being really identical to that of Dyson. What is more, Oguchi has demonstrated an equivalence between the Holstein–Primakoff representation and Dyson's method. On the other hand, Tahir-Kheli and Ter Haar[21] used the Green function method. (As we know this method consists in studying a suitable time-correlation function for two operators in order to calculate the equations of motion for the Green function and then through a decoupling procedure the Green function of a given order is reduced to the Green function of a smaller order.) The problem of a ferromagnet yields a veriety of ways of developing the analysis owing to a possibility of various boson representations for the spin components. In their first paper Tahir–Kheli and Ter Haar obtained a correction of the form T^3 which to the best of our knowledge is incorrect. However, in the second paper the above authors arrived at the correction having a form T^4. It is interesting to emphasize that Yakovlev[22] has independently obtained a similar result.

A simple approximation for the energy of spin waves can be obtained by neglecting fluctuations of the z-component around the equilibrium value. This is equivalent to replacing equation (3.95), or equation (3.97), with

$$E(T, \boldsymbol{q}) \simeq 2\sigma[J(0) - J(\boldsymbol{q})] \tag{3.111}$$

where

$$\sigma \equiv N^{-1} \langle S^z_{\boldsymbol{q}} \rangle$$

as $\boldsymbol{q} \to 0$. Bogoliubov and Tyablikov[23] investigated this approximation by calculating a double-time Green function. By doing so they have developed a theory of ferromagnetism which holds approximately at all temperatures. Unfortunately their result disagrees with the Dyson theory of spin waves in a low-temperature region at least. However, it has been proved by Keffer and Loudon[24] and independently by Morri and Kawasaki[25] that such a disagreement could be removed if the following fact is taken into account. A spin wave as generated in the region of long wavelength spin waves, which are present all the time, feels the exchange interaction coming from the equilibrium magnetization $\langle \boldsymbol{S}_{\boldsymbol{q}} \rangle$ rather than from a nonequilibrium magnetization which has been moved along with the existing background spin waves.

We conclude the present section by noting that a dynamical correction having the form T^4 in a low-temperature expansion for the magnetization is not observed experimentally with a sufficient certainty. On the other hand, it is possible to observe the energy variations of spin waves in the form

$$E(T, q) = 2SJ(1-\alpha\tau^{5/2})\,(qa)^2 \tag{3.112}$$

where α is a constant.† This temperature variation is measured in Ni up to the transition point and a very nice agreement is reached up to the temperature $T/T_c \simeq 0.75$, see Marshall.[26] It seems therefore that the law $T^{5/2}$ can be deduced from a rather general consideration.

3.7. The dispersion law for antiferromagnets with a double Heisenberg exchange interaction

The problem of the determination of the dispersion law for antiferromagnets is a natural generalization of the similar problem in ferromagnets. The only difference concerning the crystal structure is that here we introduce two identical sublattices which interpenetrate one into another in a certain way. It is of interest to quote a number of references dealing with a similar problem in a contemporary literature, namely Kaplan,[27] Kittel,[28] Saenz,[29] Liu,[30] Nagai,[31] and others. Examples of antiferromagnetic media include manganous oxide and similar substances.

We consider in the present section a simplified model of antiferromagnets in order to formulate the elementary excitations. Suppose that the crystal structure consists of two identical sublattices A and B (see Fig. 3.1) in such a way as to represent the ferromagnetic spins sitting at the A sites antiparallel to the ferromagnetic spins sitting at the B sites. For the reason of simplicity we assume that both ferro-

† Having used equation (3.105) we obtain the exact numerical value for α,

$$\alpha = \frac{0.015}{2S}.$$

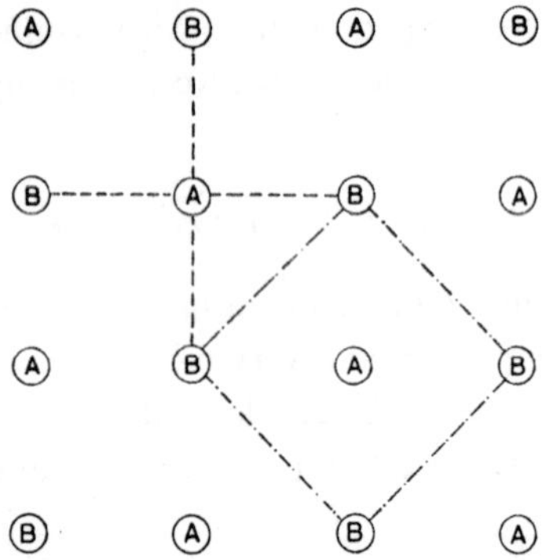

FIG. 3.1. A sequence of ferromagnetic spins in an antiferromagnet consisting of two identical interpenetrating sublattices A and B. The broken lines indicate the nearest-neighbor interactions of the magnitude J_{AB}; the dot-dash-dot lines indicate next nearest-neighbor interactions of the magnitude $J_{AA} = J_{BB}$.

magnetic spins have the same magnitude S, and that an exchange interaction of the Heisenberg type acts between them, in particular

$$\mathcal{H} = \sum_{\langle jk\rangle} [J_{Ajk}(\mathbf{S}_{Aj}\cdot\mathbf{S}_{Ak})+J_{Bjk}(\mathbf{S}_{Bj}\cdot\mathbf{S}_{Bk})] + \sum_{\langle jk\rangle} J_{ABjk}(\mathbf{S}_{Aj}\cdot\mathbf{S}_{Bk})-2g\mu_B H \sum_j (S_{Azj}-S_{Bzj}) \quad (3.113)$$

where the summation is extended over all lattice sites. The last term represents the effect of an external magnetic field. In the following analysis we assume

$$\sum_{\langle jk\rangle} J_{Ajk} = \sum_{\langle jk\rangle} J_{Bjk} = z_{AA}J_{AA}N,$$
$$\sum_{\langle jk\rangle} J_{ABjk} = z_{AB}J_{AB}N \quad (3.114)$$

where z_{AB} and z_{AA} designate the number of nearest and next-nearest neighbors, respectively. Here N is the total number of sublattice sites. (Therefore the total number of sites on the entire lattice is twice as large, $2N$.)

To expose the dynamical behavior of the ferromagnetic spins in the ground state we use a Dyson–Maleev representation which consists of replacing the spin components with two pairs of boson creation and annihilation operators as follows:

$$\begin{aligned} S_{A(z)j} &= S - n_j, \\ S_{A(+)j} &= \sqrt{(2S)}\, a_j, \\ S_{A(-)j} &= \sqrt{(2S)}\, a_j^\dagger (1 - n_j/2S), \\ n_j &= a_j^\dagger a_j; \end{aligned} \tag{3.115}$$

$$\begin{aligned} S_{B(z)j} &= -S + m_j, \\ S_{B(+)j} &= \sqrt{(2S)}\, b_j^\dagger, \\ S_{B(-)j} &= \sqrt{(2S)}\,(1 - m_j/2S)\, b_j, \\ m_j &= b_j^\dagger b_j. \end{aligned} \tag{3.116}$$

The above operators satisfy the commutation relations for the spin components exactly. Clearly we have

$$[a_j, a_k^\dagger]_{(-)} = [b_j, b_k^\dagger]_{(-)} = \Delta(j-k), \tag{3.117}$$

whereas the operators a_j and b_k commute one with another in all combinations. Next we introduce the spin-wave variable through the boson creation and annihilation operators c_q and d_q as follows:

$$\begin{aligned} c_q &= N^{-1/2} \sum_j \exp(i\boldsymbol{q}\cdot\boldsymbol{R}_j)\, a_j, \\ d_q &= N^{-1/2} \sum_j \exp(-i\boldsymbol{q}\cdot\boldsymbol{R}_j)\, b_j. \end{aligned} \tag{3.118}$$

A similar definition holds for the corresponding creation operators $c_q^\dagger$ and $d_q^\dagger$. Here $\boldsymbol{q}$ denotes a reciprocal lattice vector, while $\boldsymbol{R}_j$ stands for the position vector of a given lattice site in ordinary space.

Using the transformation (3.115–3.116) we can write the hamiltonian (3.113) in the form

$$\mathscr{H} = E_0 + \mathscr{H}_{(2)} + \mathscr{H}_{(4)} + \mathscr{H}_{(6)} \tag{3.119}$$

where

$$\begin{aligned} E_0 &= -NS(-2z_{AA}J_{AA}S + 2z_{AB}J_{AB}S + 4g\mu_B H), \\ \mathscr{H}_{(2)} &= -S \sum_{\langle jl \rangle} J_{Ajl}(n_j + n_l - a_j^\dagger a_l - a_j a_l^\dagger) \\ &\quad - S \sum_{\langle jl \rangle} J_{Bjl}(m_j + m_l - b_j^\dagger b_l - b_j b_l^\dagger) \\ &\quad + S \sum_{\langle jl \rangle} J_{ABjl}(n_j + m_l + a_j^\dagger b_l^\dagger + a_j b_l) \\ &\quad + 2g\mu_B H \sum_j (n_j + m_j), \end{aligned} \tag{3.120}$$

$$\mathscr{H}_{(4)} = -\sum_{\langle jl \rangle}\left\{J_{Ajl}\left(\frac{a_j^\dagger a_j^2 a_l^\dagger + a_j^\dagger a_l^\dagger a_l^2}{2} - n_j n_l\right)\right.$$
$$+ J_{ABjl}\left(\frac{a_j^\dagger a_j^2 b_l + a_j b_l^\dagger b_l^2}{2} + n_j m_l\right)$$
$$\left.+ J_{Bjl}\left(\frac{b_j^\dagger b_j^2 b_l^\dagger + b_j^\dagger b_l^\dagger b_l^2}{2} - m_j m_l\right)\right\} \tag{3.121}$$

$$\mathscr{H}_{(6)} = (2S)^{-1}\sum_{\langle jl \rangle} J_{ABjl}\, a_j^\dagger a_j^2 b_l^\dagger b_l^2. \tag{3.122}$$

The term $\mathscr{H}_{(6)}$ is the only nonhermitian part of the total hamiltonian. In the next transformations we shall neglect this nonhermitian part and will pay attention to only the terms $\mathscr{H}_{(2)}$ and $\mathscr{H}_{(4)}$: $\mathscr{H}_{(2)}$ will be considered in detail in the present section, whereas $\mathscr{H}_{(4)}$ will be considered in Chapter 4.

Using the Fourier transform (3.118) we arrive at

$$\mathscr{H}_{(2)} = \sum_{q} \lambda(q)(c_q^\dagger c_q + d_q^\dagger d_q)$$
$$+ \sum_{q} \mu(q)(c_q^\dagger d_q^\dagger + c_q d_q), \tag{3.123}$$

where

$$\lambda(q) = 2S[z_{AA}J_{AA} - J_{AA}(q)] + 2Sz_{AB}J_{AB} + 2g\mu_B H, \tag{3.124a}$$
$$\mu(q) = 2SJ_{AB}(q), \tag{3.124b}$$
$$J_{AA}(q) = \sum J_{Ajl}\exp[iq\cdot(R_j - R_l)]$$
$$J_{AB}(q) = \sum J_{ABjl}\exp[iq\cdot(R_j - R_l)]. \tag{3.125}$$

In the following analysis we diagonalize the above hamiltonian by introducing a canonical transformation in the same form as applied by Bogoliubov[32] to the system of strongly interacting bosons. The transformation in question consists of replacing the operators c_q and d_q by new operators A_q and B_q in the form

$$A_q = u(q)\, c_q - v(q)\, d_q^\dagger,$$
$$A_q^\dagger = u(q)\, c_q^\dagger - v(q)\, d_q, \tag{3.126a}$$
$$B_q = u(q)\, d_q - v(q)\, c_q^\dagger,$$
$$B_q^\dagger = u(q)\, d_q^\dagger - v(q)\, c_q. \tag{3.126b}$$

The following commutation relations hold

$$[A_q, A^\dagger_{q'}]_{(-)} = [B_q, B^\dagger_{q'}]_{(-)} = \Delta(q-q'). \tag{3.127}$$

Clearly, any other commutator involving either two creation or two annihilation operators of the same type, or any pair of operators of different types, is equal to zero for all values of q and q'. The coefficients $u(q)$ and $v(q)$ are real and even functions of q; they satisfy the condition

$$|u(q)|^2 - |v(q)|^2 = 1. \tag{3.128}$$

Let us write the hamiltonian $\mathcal{H}_{(2)}$ in the form

$$\mathcal{H}_{(2)} = \sum E(q)(A^\dagger_q A_q + B^\dagger_q B_q), \quad E(q) = \hbar\omega(q) \tag{3.129}$$

where $\omega(q)$ denotes the frequency of normal vibrations. The equations of motion read

$$\begin{aligned}[\mathcal{H}_{(2)}, A^\dagger_q] &= E(q)\, A^\dagger_q,\\ [\mathcal{H}_{(2)}, A_q] &= -E(q)\, A_q,\end{aligned} \tag{3.130a}$$

$$\begin{aligned}[\mathcal{H}_{(2)}, B^\dagger_q] &= E(q)\, B^\dagger_q\\ [\mathcal{H}_{(2)}, B_q] &= -E(q)\, B_q.\end{aligned} \tag{3.130b}$$

On the other hand, we may write

$$\begin{aligned}[\mathcal{H}_{(2)}, A^\dagger_q] &= \lambda(q)[u(q)\, c^\dagger_q + v(q)\, d_q]\\ &\quad + \mu(q)[u(q)\, d_q + v(q)\, c^\dagger_q]\\ &= E(q)[u(q)\, c^\dagger_q - v(q)\, d_q],\end{aligned} \tag{3.131a}$$

$$\begin{aligned}[\mathcal{H}_{(2)}, A_q] &= -\lambda(q)[u(q)\, c_q + v(q)\, d^\dagger_q]\\ &\quad - \mu(q)[u(q)\, d^\dagger_q + v(q)\, c_q]\\ &= -E(q)[u(q)\, c_q - v(q)\, d^\dagger_q].\end{aligned} \tag{3.131b}$$

By comparing the coefficients in front of the operators c_q and d_q in (3.131a) we obtain two equations for the determination of $u(q)$, $v(q)$, and $E(q)$

$$\begin{aligned}[\lambda(q)-E(q)]\, u(q) + \mu(q)\, v(q) &= 0,\\ \mu(q)\, u(q) + [\lambda(q)+E(q)]\, v(q) &= 0.\end{aligned} \tag{3.132}$$

Had we started from equation (3.131b) we would have obtained the same set of equations. The solution to equation (3.132) is given by

$$E^2(\boldsymbol{q}) = \lambda^2(\boldsymbol{q}) - \mu^2(\boldsymbol{q}), \tag{3.133a}$$

$$u(\boldsymbol{q}) = \left[\frac{\lambda(\boldsymbol{q})+E(\boldsymbol{q})}{2E(\boldsymbol{q})}\right]^{1/2}$$

$$v(\boldsymbol{q}) = -\left[\frac{\lambda(\boldsymbol{q})-E(\boldsymbol{q})}{2E(\boldsymbol{q})}\right]^{1/2}. \tag{3.133b}$$

Let us now investigate the behavior of $E(\boldsymbol{q})$ for small values of the reciprocal lattice vector $\boldsymbol{q}$. With the help of the expressions

$$J_{AA}(\boldsymbol{q}) = z_{AA}J_{AA}[1-\tfrac{1}{6}(qb_{AA})^2],$$
$$J_{AB}(\boldsymbol{q}) = z_{AB}J_{AB}[1-\tfrac{1}{6}(qb_{AB})^2], \tag{3.134}$$

we obtain

$$E^2(\boldsymbol{q}) = P + Qq^2 \tag{3.135a}$$

which for a weak magnetic field H reduces to

$$P = 8Sz_{AB}J_{AB}g\mu_B H,$$
$$Q = \tfrac{4}{3}[(Sz_{AB}J_{AB}b_{AB})^2 - Sz_{AA}J_{AA}(Sz_{AB}J_{AB} - g\mu_B H)\, b_{AA}^2]. \tag{3.135b}$$

Here b_{AA} denotes the distance separating two nearest ferromagnetic spins situated at the sublattice A, whereas b_{AB} denotes the distance separating a ferromagnetic spin situated at the sublattice A from the nearest antiferromagnetic spins situated at the sublattice B, both having a cubic symmetry.

It is of interest to consider a special case where $H = 0$. Now the above equations lead to

$$P = 0,$$
$$E(\boldsymbol{q}) = c\hbar q \tag{3.136a}$$

where the quantity c has the value

$$c = \frac{2S}{\sqrt{(3)}\cdot\hbar}[z_{AB}J_{AB}(z_{AB}J_{AB}b_{AB}^2 - z_{AA}J_{AA}b_{AA}^2)]^{1/2} \tag{3.136b}$$

and can be interpreted as the velocity of antiferromagnetic magnons.

A number of lattice and dynamical parameters are evaluated for antiferromagnetic substances with the NaCl lattice structure (cubic F) and collected in Table 3.7 and Table 3.8. Here we set up

$$J_{AB} = \tfrac{1}{2}J_1 + 2J_2, \quad J_{AA} = \tfrac{1}{2}J_1 \tag{3.137}$$

TABLE 3.7.

Lattice Parameters for Antiferromagnetic Substances with NaCl *Lattice Structure (Cubic F)*

Substance	Nearest-neighbor distance b_{AB}, Å	Next-nearest-neighbor distance b_{AA}, Å
MnO	2.22‡	3.14§
FeO	2.16	3.06
CoO	2.13	3.01
NiO	2.09	2.96
MnS	2.61	3.69

‡ After Slater.(14) § Assuming $b_{AA} = \sqrt{(2)} \cdot b_{AB}$.

TABLE 3.8.

Pseudo-spin (S), *Transition Temperature* (T_c), *Néel Temperature* (θ), *Exchange Integrals* (J_1, J_2), *and Velocity of Magnons* (c) *for Antiferromagnetic Substances with* NaCl *Lattice Structure (Cubic F)*

Substance	S	T_c, K	$-\theta$, K	$-J_1$, K	$-J_2$, K	c, 10^5 cm/sec
MnO	$\frac{5}{2}$	116	610	7.2‡	3.5‡	imaginary
				14.0§	14.0§	7.92
				3.83§	3.37§	5.99
FeO	2	198	570	7.8‡	8.2‡	3.84
CoO	$\frac{3}{2}$	292	330	1.0‡	20‡	11.48
				6.9‡	21.6‡	9.88
				1.01‖	16.64‖	11.43
NiO	1	523	3000	150‡	95‡	imaginary
			1310	50‡, §	85‡, §	25.90
α-MnS	$\frac{5}{2}$	154	465	4.4‡, §	4.5‡, §	3.08
				7§	12.5§	4.74
β-MnS	$\frac{5}{2}$	155	982	10.5‡	7.2‡	imaginary

‡ After Smart.(33) § After Reissland and Begum.(34) ‖ After Buyers *et al.*(35)

in order to evaluate the velocity of antiferromagnetic magnons. The condition for this quantity to be a real magnitude is seen from equation (3.136b). Writing this equation in a slightly different form we obtain

$$c = \frac{12S\,|J_{\mathrm{AB}}|\,b_{\mathrm{AB}}}{\sqrt{(3)}\cdot\hbar}\left(1-4\,\frac{J_{\mathrm{AA}}}{J_{\mathrm{AB}}}\right)^{1/2}. \tag{3.136c}$$

If

$$|J_{\mathrm{AB}}| > 4\,|J_{\mathrm{AA}}| \tag{3.138}$$

then the velocity of antiferromagnetic magnons is real. When this condition is not fulfilled the result is marked by "imaginary" (see Table 3.8).

3.8. The high-temperature static susceptibility, I

At high temperatures we do not expect magnetic elementary excitations of the kind so far considered to be present in a magnetic system. A predominant phenomenon taking place is the exchange process whose duration is measured by the exchange interval,

$$\tau_{\mathrm{exch}} \simeq 10^{-14}\ \mathrm{sec}.$$

On a macroscopic scale this proves an extremely short time so that all physical quantities are evaluated as thermal averages over the statistically independent exchange states. The present problem is approached by two simple and yet very successful approximations.

At first we may consider one of the ferromagnetic spins in the exchange hamiltonian in order to replace its interaction with the neighboring spins by an effective field over the direct lattice space. This is physically equivalent as if we had assumed that each spin is statistically independent. The present method is named the molecular field approximation.

Another approach consists of considering the Fourier components of one ferromagnetic spin independently and replace its interaction by an effective field over the reciprocal lattice space. This method is called the random phase approximation.

Using the latter method we can write[36]

$$\mathcal{H}_{\mathrm{RPA}} = -2\sum_{q} [J(-\boldsymbol{q})\, \boldsymbol{S}_q \cdot \langle \boldsymbol{S}_{-q}\rangle + J(\boldsymbol{q})\langle \boldsymbol{S}_q\rangle \cdot \boldsymbol{S}_{-q}] \qquad (3.139)$$

where summation is extended over the first Brillouin zone at least. Here

$$\boldsymbol{S}_q = N^{-1/2}\sum_f \exp\,(i\boldsymbol{q}\cdot\boldsymbol{R}_f)\,\boldsymbol{S}_f. \qquad (3.140a)$$

At high temperatures the only component which does not vanish identically is that one parallel to the applied field,

$$\langle \boldsymbol{S}_{-q}\rangle = \{0, 0, \langle S_q^z\rangle\}, \qquad (3.140b)$$

$$\begin{aligned}\mathcal{H}_{\mathrm{RPA}} = &-2[J(-\boldsymbol{q})\langle S_q^z\rangle\, S_{-q}^z + J(q)\langle S_q^z\rangle\, S_{-q}^z] \\ &+ g\mu_{\mathrm{B}} H_{\mathrm{app}}(S_q^z + S_{-q}^z) = \boldsymbol{H}_{\mathrm{eff}}(-\boldsymbol{q})\cdot\boldsymbol{S}_q. \end{aligned} \qquad (3.141)$$

Here the total effective field which a particular Fourier component feels is defined by

$$\boldsymbol{H}_{\mathrm{eff}}(-\boldsymbol{q}) = \{0, 0, H_{\mathrm{eff}}(-\boldsymbol{q})\}, \qquad (3.142a)$$

$$H_{\mathrm{eff}}(-\boldsymbol{q}) = -2\,\frac{J(-\boldsymbol{q})}{g\mu_{\mathrm{B}}}\,\langle S_{-q}^z\rangle + H_{\mathrm{app}}. \qquad (3.142b)$$

At high temperatures a paramagnetic static susceptibility is determined by the Curie law,

$$\chi = \frac{C}{T}$$

with

$$C = N\,\frac{(g\mu_{\mathrm{B}})^2\, S(S+1)}{3k}. \qquad (3.143)$$

The Curie law will be altered in the presence of an exchange interaction. To see this alteration we introduce the response to the applied field for each Fourier component independently through the magnetization of the system. Hence

$$\boldsymbol{M}(\boldsymbol{q}) = \hat{\chi}(\boldsymbol{q}) \otimes \boldsymbol{H}_{\text{eff}}(\boldsymbol{q}) \tag{3.144}$$

where $\hat{\chi}(\boldsymbol{q})$ is a tensor quantity, the symbol $\otimes$ is used to indicate the tensor multiplication. In the present approximation

$$\boldsymbol{M}(\boldsymbol{q}) = \{0, 0, M_z(\boldsymbol{q})\}$$

$$\hat{\chi}(\boldsymbol{q}) = \begin{bmatrix} \chi_x(\boldsymbol{q}) & 0 & 0 \\ 0 & \chi_y(\boldsymbol{q}) & 0 \\ 0 & 0 & \chi_z(\boldsymbol{q}) \end{bmatrix}. \tag{3.145}$$

Also we have

$$M_z(\boldsymbol{q}) = -g\mu_{\text{B}}\langle S_{\boldsymbol{q}}^z\rangle = -2\frac{J(\boldsymbol{q})\,\chi_z(\boldsymbol{q})}{g\mu_{\text{B}}}\langle S_{\boldsymbol{q}}^z\rangle + \chi_z(\boldsymbol{q})\,H_{\text{app}}. \tag{3.146}$$

There follows the solution for a z-component of the spin.

$$-g\mu_{\text{B}}\langle S_{\boldsymbol{q}}^z\rangle = \frac{\chi_z(\boldsymbol{q})}{1-2\dfrac{J(\boldsymbol{q})}{(g\mu_{\text{B}})^2}\chi_z(\boldsymbol{q})}H_{\text{app}}.$$

The next step is to identify the high-temperature static susceptibility with a Curie law,

$$\lim_{T\to\infty} \chi_z(\boldsymbol{q}) = \frac{C}{T},$$

which leads to

$$\chi_z(\boldsymbol{q}) = \frac{C}{T-2CJ(\boldsymbol{q})/(g\mu_{\text{B}})^2}. \tag{3.147}$$

There are two important phenomena that immediately follow from equation (3.147).

1. Ferromagnetism occurs when the exchange interaction is essentially positive over the first Brillouin zone at least. In this case the transition temperature is defined by

$$T_c = 2C\frac{J(0)}{(g\mu_{\text{B}})^2}$$

where $J(0)$ is the maximum value of the exchange interaction. (This is true for the first Brillouin zone at least.) Since the susceptibility is increased infinitely for this particular value the corresponding Fourier component of the magnetization remains finite even if the magnitude of the applied field is equal to zero. By inserting $J(0) = zJ$ in the above equation we obtain

$$T_c\,(\text{ferro}) = \frac{2zJS(S+1)}{3k} \tag{3.148a}$$

in agreement with the molecular field approximation. Therefore the static susceptibility becomes

$$\chi(0)\,(\text{ferro}) = \frac{C}{T-T_c\,(\text{ferro})}\,. \tag{3.148b}$$

2. Antiferromagnetism occurs when the exchange interaction is essentially negative over the first Brillouin zone at least. This will happen at the edge of the zone for which

$$\boldsymbol{q} = \pm\pi(\boldsymbol{a}^* + \boldsymbol{b}^* + \boldsymbol{c}^*)$$

where $\boldsymbol{a}^*$, $\boldsymbol{b}^*$, and $\boldsymbol{c}^*$ designate the basic reciprocal lattice vectors of elementary translations. Here one obtains

$$J(\boldsymbol{q}) = -J(0) = -zJ = z\,|J|$$

as J is essentially negative. Hence the transition temperature becomes

$$T_c\,(\text{antif}) = \frac{2z\,|J|\,S(S+1)}{3k} \tag{3.149a}$$

whereas the static susceptibility becomes

$$\chi(0)\,(\text{antif}) = \frac{C}{T+T_c\,(\text{antif})}\,. \tag{3.149b}$$

This behavior is identical with the Néel law of antiferromagnetism, where the only difference is that the Néel temperature is replaced by T_c (antif).

3.9. The high-temperature static susceptibility, II

Ferrimagnetism occurs in those lattices which are composed of non-equivalent magnetic ion sublattices. The exchange hamiltonian in this case is identical to that introduced in Section 3.7 except that in general S_A may be different from S_B. A random-phase approximation holds formally invariant with respect to an instantaneous interchange of all quantities referring to the sublattice A with the corresponding quantities referring to the sublattice B. Let us consider the latter one. We write

$$\begin{aligned}\mathscr{H}_{\mathrm{RPA}} = 2 \sum_{A \neq B} \{ [J_{AB}(-\boldsymbol{q})\langle S^z_{A,-\boldsymbol{q}}\rangle + J_{BB}(-\boldsymbol{q})\langle S^z_{B,-\boldsymbol{q}}\rangle]\, S^z_{B,\boldsymbol{q}} \\ + [J_{AB}(\boldsymbol{q})\langle S^z_{A,\boldsymbol{q}}\rangle + J_{BB}(\boldsymbol{q})\langle S^z_{B,\boldsymbol{q}}\rangle]\, S^z_{B,-\boldsymbol{q}} \} \\ - g\mu_B H_{\mathrm{app}}(S^z_{B,\boldsymbol{q}} + S^z_{B,-\boldsymbol{q}})\end{aligned} \tag{3.150}$$

to show that the motion of a given spin on the sublattice B is determined by a thermal average taken over all neighbor spins on both sublattices. The introduced Fourier transform depends on a particular lattice structure under consideration.

By definition the magnetization on the sublattice B,

$$\begin{aligned}M_{B,z}(\boldsymbol{q}) = -g\mu_B\langle S^z_{B,\boldsymbol{q}}\rangle = \frac{2}{g\mu_B}[J_{AB}(\boldsymbol{q})\langle S^z_{A,\boldsymbol{q}}\rangle + J_{BB}(\boldsymbol{q})\langle S^z_{B,\boldsymbol{q}}\rangle] \\ \times \chi_{B,z}(\boldsymbol{q}) - \chi_{B,z}(\boldsymbol{q})\, H_{\mathrm{app}}\end{aligned} \tag{3.151a}$$

whereas on the sublattice A,

$$\begin{aligned}M_{A,z}(\boldsymbol{q}) = -g\mu_B\langle S^z_{A,\boldsymbol{q}}\rangle = \frac{2}{g\mu_B}[J_{AB}(\boldsymbol{q})\langle S^z_{B,\boldsymbol{q}}\rangle + J_{AA}(\boldsymbol{q})\langle S^z_{A,\boldsymbol{q}}\rangle] \\ \times \chi_{A,z}(\boldsymbol{q}) + \chi_{A,z}(\boldsymbol{q})\, H_{\mathrm{app}}.\end{aligned} \tag{3.151b}$$

The solution to the above two equations for the z-component of the spin is given by

$$\langle S^z_{A,\boldsymbol{q}}\rangle = \frac{\Delta_1}{\Delta}, \quad \langle S^z_{B,\boldsymbol{q}}\rangle = \frac{\Delta_2}{\Delta}$$

where

$$\Delta_1 = \frac{1}{f}[2J_{AB}(\boldsymbol{q})\,\chi_{A,z}(\boldsymbol{q})\,\chi_{B,z}(\boldsymbol{q})$$

$$+2J_{BB}(\boldsymbol{q})\,\chi_{A,z}(\boldsymbol{q})\,\chi_{B,z}(\boldsymbol{q})+f^2\,\chi_{A,z}(\boldsymbol{q})]\,H_{app} \qquad (3.152a)$$

$$\Delta_2 = -\frac{1}{f}[2J_{AB}(\boldsymbol{q})\,\chi_{A,z}(\boldsymbol{q})\,\chi_{B,z}(\boldsymbol{q})$$

$$+2J_{AA}(\boldsymbol{q})\,\chi_{A,z}(\boldsymbol{q})\,\chi_{B,z}(\boldsymbol{q})+f^2\,\chi_{B,z}(\boldsymbol{q})]\,H_{app}, \qquad (3.152b)$$

$$\Delta = \frac{1}{f^2}\{4[J_{AB}^2(\boldsymbol{q})-J_{AA}(\boldsymbol{q})\,J_{BB}(\boldsymbol{q})]\,\chi_{A,z}(\boldsymbol{q})\,\chi_{B,z}(\boldsymbol{q})$$

$$-2[J_{AA}(\boldsymbol{q})\,\chi_{A,z}(\boldsymbol{q})+J_{BB}(\boldsymbol{q})\,\chi_{B,z}(\boldsymbol{q})]\,f^2-f^4\}, \qquad (3.152c)$$

$$f = g\mu_B.$$

The magnetization and static susceptibility on the entire lattic become

$$M_z(\boldsymbol{q}) = -g\mu_B\frac{\Delta_1-\Delta_2}{\Delta},$$

$$\chi_z(\boldsymbol{q}) = \frac{M_z(\boldsymbol{q})}{H_{app}}. \qquad (3.153)$$

Using equations (3.151–3.153) we arrive at the following result:

$$\frac{1}{\chi_z(\boldsymbol{q})} = \frac{T}{C}+\frac{1}{\chi_0}-\frac{\sigma}{T-\theta} \qquad (3.154a)$$

with

$$C = C_A+C_B,$$

$$\theta = -\frac{4J_{AB}(\boldsymbol{q})\,2J_{AA}(\boldsymbol{q})+2J_{BB}(\boldsymbol{q})}{(g\mu_B)^2\,C}\,C_AC_B,$$

$$\frac{1}{\chi_0} = 2\frac{J_{AA}(\boldsymbol{q})\,C_A^2+J_{BB}(\boldsymbol{q})\,C_B^2-2J_{AB}(\boldsymbol{q})\,C_AC_B}{(g\mu_B)^2\,C^2},$$

$$\sigma = \frac{4}{(g\mu_B)^4}\Big[(2J_{AB}(\boldsymbol{q})+J_{AA}(\boldsymbol{q})+J_{BB}(\boldsymbol{q}))$$

$$\times(J_{AA}(\boldsymbol{q})\,C_A^2+J_{BB}(\boldsymbol{q})\,C_B^2-2J_{AB}(\boldsymbol{q})\,C_AC_B)\,\frac{C_AC_B}{C^3}$$

$$+(J_{AB}^2(\boldsymbol{q})-J_{AA}(\boldsymbol{q})\,J_{BB}(\boldsymbol{q}))\,\frac{C_AC_B}{C}\Big] \qquad (3.154b)$$

where we have set up, in analogy with ferromagnetism,

$$\lim_{T\to\infty} \chi_{\mathrm{A},z}(\boldsymbol{q}) = \frac{C_{\mathrm{A}}}{T},$$

$$\lim_{T\to\infty} \chi_{\mathrm{B},z}(\boldsymbol{q}) = \frac{C_{\mathrm{B}}}{T}. \tag{3.154c}$$

The introduced Curie constants are given by

$$C_{\mathrm{A}} = N_{\mathrm{A}}(g\mu_{\mathrm{B}})^2 S_{\mathrm{A}}(S+1)/3k,$$

$$C_{\mathrm{B}} = N_{\mathrm{B}}(g\mu_{\mathrm{B}})^2 S_{\mathrm{B}}(S_{\mathrm{B}}+1)/3k. \tag{3.154d}$$

The expressions (3.154) are in a good agreement with the Néel law of ferrimagnetism as introduced in Section 2.2.

In practical work the Néel law can be greatly simplified if certain assumptions on the dynamical basis are made, in particular:

(a) consider the centre of the first Brillouin zone, and
(b) neglect all the interactions except those among the nearest neighbors.

These assumptions lead to the following set of Néel parameters:

$$\theta = \frac{4z_{\mathrm{AB}}\,|J_{\mathrm{AB}}|\,C_{\mathrm{A}}C_{\mathrm{B}}}{(g\mu_{\mathrm{B}})^2\,C},$$

$$\frac{1}{\chi_0} = \frac{\theta}{C},$$

$$\sigma = \frac{C(C_{\mathrm{A}}-C_{\mathrm{B}})^2}{4C_{\mathrm{A}}C_{\mathrm{B}}}\,\theta^2, \tag{3.155}$$

which lead to the static susceptibility

$$\chi = \frac{C(T-\theta)}{T^2-T_c^2\,(\text{ferri})}$$

with

$$T_c\,(\text{ferri}) = \theta\left[1+\frac{C^2(C_{\mathrm{A}}-C_{\mathrm{B}})^2}{4C_{\mathrm{A}}C_{\mathrm{B}}}\right]^{1/2}. \tag{3.156}$$

TABLE 3.9.

Number of Bohr Magnetons (n_B), *Pseudo-spin* (S_M), *Curie Constants* (C_A, C_B), *Curie Temperature* (θ), *and Transition Temperature* (T_c) *of Ferrites*

Substance	n_B‡ / molecule	S_M	C_A cm^{-3} erg K/gauss^2	C_B cm^{-3} erg K/gauss^2	θ,‡ K	T_c, K
$MnFe_2O_4$	4.4–5.0	Mn : 5/2	6.49×10^{-2}	22.24×10^{-2}	573	583
Fe_3O_4	4.0–4.1	Fe : 2	9.35	26.46	858	874
$CoFe_2O_4$	3.7–3.9	Co : 2	7.80	22.05	793	803
$NiFe_2O_4$	2.2–2.4	Ni : 1	10.60	19.09	858	861
$CuFe_2O_4$	1.3–2.3	Cu : 1	5.30	9.54	728	728.7
$MgFe_2O_4$	0.9–1.4	Mg : 1/2	8.73	11.74	713	713.3

‡ After Bozorth.[37]

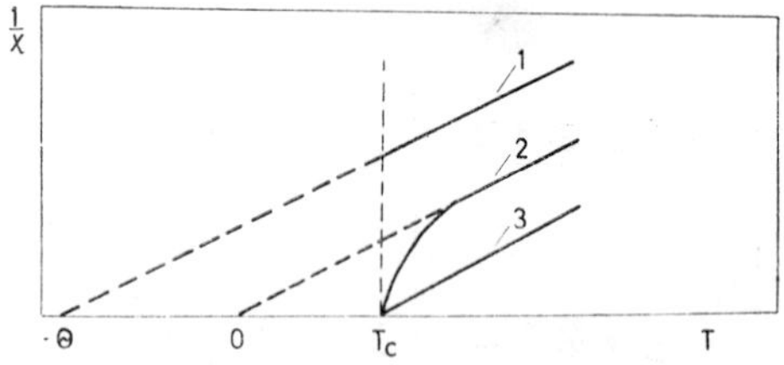

FIG. 3.2. The high-temperature inverse static susceptibility calculated for various magnetic systems: (1) antiferromagnet; (2) ferrimagnet; (3) ferromagnet.

The value of an inverse susceptibility for various cases is illustrated in Fig. 3.2. The present method is also applied to a number of known ferrites, see Table 3.9, assuming

$$S_A = \tfrac{5}{2},$$
$$S_B = S_A + S_M,$$
$$C_A = \frac{M_s g \mu_B}{3kS_M} S_A(S_A+1),$$
$$C_B = C_A \frac{S_B(S_B+1)}{S_A(S_A+1)}, \tag{3.157}$$

where S_M and M_s designate the pseudo-spin of a divalent metal of the compound $MO \cdot Fe_2O_3$ and the observed saturated magnetization, respectively. The difference T_c(ferri)-θ depends on the used data and varies from several parts of a degree up to 16 K at most (Fe_3O_4).

References

1. W. Pauli, *Phys. Rev.* **58,** 716 (1940).
2. A. I. Akhiezer and V. B. Berestetskii, *Quantum Electrodynamics*, sect. 14, Interscience Publishers, Ltd., New York, 1965.
3. J. C. Slater, *Quantum Theory of Matter*, McGraw-Hill Book Co., Inc., New York, 1951.
4. J. H. Van Vleck, *Phys. Rev.* **49,** 232 (1936).
5. J. H. Van Vleck, *Rev. Mod. Phys.* **17,** 27 (1945); *ibid.* **25,** 220 (1953).
6. A. Herpin, *Théorie du magnétisme*, chapts. II, VII, and IX, INSTN, Saclay, PUF, 108, Boulevard Saint-Germain, Paris, 1968.

7. A. R. EDMONDS, *Angular Momentum in Quantum Mechanics*, chap. 2, Princeton University Press, Princeton, N. J., 1957.
8. T. HOLSTEIN and H. PRIMAKOFF, *Phys. Rev.* **58,** 1098 (1940).
9. P. JORDAN and E. WIGNER, *Z. Physik* **47,** 631 (1928) [also in: *Selected Papers on Quantum Electrodynamics*, p. 41, ed. by J. SCHWINGER, Dover Publications, Inc., New York, 1958].
10. N. N. BOGOLIUBOV and D. V. SHIRKOV, *Introduction to the Theory of Quantized Fields*, sect. 3, Interscience Publishers, Ltd., New York, 1959.
11. C. KITTEL, *Introduction to Solid State Physics*, chap. 15, John Wiley & Sons, Inc., New York, 1966.
12. R. M. BOZORTH, *Ferromagnetism*, D. Van Nostrand Co., Inc., New York, 1951.
13. W. LOMER and W. MARSHALL, *Phil. Mag.* **3,** 185 (1958).
14. J. C. SLATER, *Quantum Theory of Molecules and Solids*, McGraw-Hill Book Co., Inc., New York, 1963 (Vol. I) and 1965 (Vol. II).
15. F. BLOCH, *Z. Physik* **61,** 206 (1930); *ibid.* **74,** 295 (1932).
16. F. J. DYSON, *Phys. Rev.* **102,** 1217, 1230 (1956).
17. J. VAN KRANENDONK, *Physica* **XXI**, 749, 925 (1955).
18. S. V. MALEEV, *Zh. Eksperim. i Teor. Fiz.* **33**, 1010 (1957) [English trans.: *Soviet Phys. — JETP* **6,** 776 (1958)].
19. T. MORITA, *Prog. Theor. Phys.* **20,** 614, 728 (1958).
20. T. OGUCHI, *Phys. Rev.* **117,** 117 (1960).
21. R. A. TAHIR-KHELI and D. TER HAAR, *Phys. Rev.* **127,** 88, 95 (1962).
22. E. N. YAKOVLEV, *Fiz. Tverd. Tela* **4,** 179 (1962) [English trans.: *Soviet Phys. — Solid State* **4,** 127 (1962)].
23. N. N. BOGOLIUBOV and S. V. TYABLIKOV, *Dokl. Akad. Nauk USSR* **126,** 53 (1959).
24. F. KEFFER and R. LOUDON, *J. Appl. Phys.* **32,** 2S (1961).
25. H. MORRI and K. KAWASAKI, *Prog. Theor. Phys.* **27,** 529 (1962).
26. W. MARSHALL, in: *Proceedings of the Eight International Conference on Low Temperature Physics*, Butterworths, London, 1963.
27. T. A. KAPLAN, *Phys. Rev.* **109,** 782 (1958).
28. C. KITTEL, *Phys. Rev.* **120,** 335 (1960).
29. A. W. SÁENZ, *Phys. Rev.* **125,** 1940 (1962).
30. S. H. LIU, *Phys. Rev.* **142,** 267 (1966).
31. O. NAGAI, *Phys. Rev.* **180,** 557 (1969).
32. Quoted under reference 16, chap. 1.
33. J. S. SMART, *Effective Field Theories of Magnetism*, W. B. Saunders Co., Philadelphia, 1966.
34. J. A. REISSLAND and N. A. BEGUM, *J. Phys. C (Solid St. Phys.)* **2,** 874 (1969).
35. W. J. L. BUYERS, G. DOLLING, J. SAKURAI and R. A. COWLEY, in: *Neutron Inelastic Scattering*, Vol. II, International Atomic Energy Agency, Vienna, 1968.
36. R. M. WHITE, *Quantum Theory of Magnetism*, McGraw-Hill Book Co., Inc., New York, 1970.
37. Quoted under reference 2, chap. 2.

CHAPTER 4

The Use of Green Functions

4.1. Basic definitions

In studying solid-state problems we often use the Green functions method. At present Green functions are divided into two large categories, one being based on a generalization of the correlation functions mostly used in statistical mechanics, the other being based on the calculation of the true wave function in various quantum-mechanical problems. In the present work we shall be concerned only with the former category.

A double-time temperature-dependent Green function is defined as a thermal average taken over the canonical ensemble for two operators which may represent the boson, fermion, or any arbitrary fields. The definition takes care of the distinction between the advanced (a), retarded (r), or causal (c) Green functions. There are different notations used to designate the Green function for two operators, say, A and B, at two time instants t and t'; respectively; in particular

$$G(A, B; t, t')_{(j)},$$

$$G_{AB}(t, t')_{(j)},$$

and

$$\langle\langle A(t) \mid B(t') \rangle\rangle_{(j)}$$

where $j = a, r,$ or c. In exposing the present two sections we have followed the works of Kubo,[1] Zubarev,[2] Tyablikov,[3] and Barry.[4]

Using the time-shifted Heisenberg operators corresponding to the mechanical variables (also named the Heisenberg representation) the double-time temperature-dependent Green functions are written

$$\langle\langle A(t) \mid B(t')\rangle\rangle_{(a)} = i\theta(t-t')\,\langle[A(t), B(t')]\rangle, \tag{4.1a}$$

$$\langle\langle A(t) \mid B(t')\rangle\rangle_{(r)} = -i\theta(t-t')\,\langle[A(t), B(t')]\rangle, \tag{4.1b}$$

$$\langle\langle A(t) \mid B(t')\rangle\rangle_{(c)} = -i\langle TA(t)\, B(t')\rangle. \tag{4.1c}$$

Here $\theta(t-t')$ designates the Heaviside function (also named the step function) defined by

$$\theta(x) = \begin{cases} 1, & \text{if} \quad x > 0, \\ 0, & \text{if} \quad x < 0. \end{cases} \tag{4.2}$$

The symbol $\langle \ldots \rangle$ is used to designate the canonical ensemble average,

$$\langle A \rangle = Z^{-1}\,\mathrm{Tr}\, A \exp\,(-\beta\mathcal{H}), \qquad \beta = \frac{1}{kT}, \tag{4.3}$$

Z is the canonical partition function, $\mathcal{H}$ is the total hamiltonian, $A(t)$ is the Heisenberg representation for A, in units $\hbar = 1$,

$$A(t) = \exp\,(i\mathcal{H}t)\, A \exp\,(-i\mathcal{H}t), \tag{4.4}$$

and T is the Wick time-ordering operator,

$$TA(t)\, B(t') = \theta(t-t')\, A(t)\, B(t') + \eta\,(t-t')\, B(t')\, A(t). \tag{4.5}$$

Finally, the general commutator operator appearing in equations (4.1) is defined by

$$[A, B] = AB - \eta BA \tag{4.6}$$

where $\eta = 1$ or -1 for boson or fermion fields, respectively. However, this is not the unique choice since the definition can be taken the other way round. Moreover, the operators A and B need not represent purely boson or purely fermion fields but any arbitrary fields. It is worth noting that only the retarded and causal Green functions will be used.

A given Green function when applied to the statistical equilibrium ensemble becomes a function of the time difference, $t-t'$. Indeed, using equations (4.1), (4.3), (4.4), and a cyclic property of the trace,

$$\operatorname{Tr} ABC = \operatorname{Tr} BCA = \operatorname{Tr} CAB,$$

as for the retarded Green function, we obtain

$$\begin{aligned}\langle\langle A(t) \mid B(t')\rangle\rangle_{(r)} = {} & -i\theta(t-t')\, Z^{-1}\{\operatorname{Tr} \exp(\mathscr{H}[-i(t-t')-\beta]) \\ & \times A \exp[i\mathscr{H}(t-t')]\, B - \eta \operatorname{Tr} \exp \mathscr{H}[i(t'-t)-\beta]) \\ & \times B \exp[-i\mathscr{H}(t'-t)]A\}. \end{aligned} \tag{4.7}$$

A similar proof holds for the remaining two Green functions.

In what follows we shall investigate the equations of motion in order to determine the Green functions explicitly. By definition the operators $A(t)$ and $B(t)$ satisfy the following equation of motion, in units $\hbar = 1$,

$$\frac{dA}{dt} = i[\mathscr{H}, A] \tag{4.8}$$

and similarly for the other operator. Therefore having differentiated equation (4.1b) with respect to time one has

$$\begin{aligned} i\frac{dG}{dt} &= i\frac{d}{dt}\langle\langle A(t) \mid B(t')\rangle\rangle \\ &= \frac{d\theta(t-t')}{dt}\langle[A(t), B(t')]\rangle \\ &\quad + \left\langle\left\langle i\frac{dA(t)}{dt} \,\middle|\, B(t')\right\rangle\right\rangle \end{aligned} \tag{4.9}$$

where the suffix (r) has been suppressed. An integral representation for the Heaviside function is used,

$$\theta(t) = \int_{-\infty}^{t} \delta(x)\, dx, \tag{4.10}$$

to write the retarded double-time temperature-dependent Green function in the form

$$i\frac{dG}{dt} = \delta(t-t')\langle[A(t), B(t')]\rangle$$

$$\langle\langle\{A(t)\mathcal{H}(t)-\mathcal{H}(t)A(t)\} \mid B(t')\rangle\rangle. \qquad (4.11)$$

In general the Green function standing on the right side of equation. (4.11) has a higher order than the original one and can be reduced in most cases of interest to the original one by a suitable decoupling procedure. This will be illustrated later.

As the Green functions depend only on the time difference, $t-t'$, it is therefore desirable and advantageous to introduce the energy representation by taking the Fourier transform over the entire time axis. Writing

$$G(A, B; t-t') = \int_{-\infty}^{\infty} G(A, B; E)\exp[-iE(t-t')]dE \quad (4.12a)$$

where E designates the energy the introduced Green function in the energy representation is equal to

$$G(A, B; E) = \frac{1}{2\pi}\int_{-\infty}^{\infty} G(A, B; \tau)\exp(iE\tau)\,d\tau. \qquad (4.12b)$$

Now having multiplied equation (4.11) with

$$\exp(iE\tau)/(2\pi)$$

in order to integrate over the entire time axis one arrives at

$$EG(A, B; E) \equiv E\langle\langle A \mid B\rangle\rangle = \frac{1}{2\pi}[A, B]+\langle\langle[A, \mathcal{H}] \mid B\rangle\rangle \quad (4.13)$$

where the following boundary condition is imposed

$$G(A, B; \tau = -\infty) = G(A, B; \tau = \infty) = 0.$$

4.2. Perturbation expansion for the Green functions

To determine the desired Green function we consider the chain of equations (4.13). The total hamiltonian according to Tyablikov[3] can be separated into a harmonic part, $\mathcal{H}_{(2)}$, and an anharmonic part, $\mathcal{H}_{A}$, as follows:

$$\mathcal{H} = \mathcal{H}_{(2)} + \varepsilon \mathcal{H}_{A} \tag{4.14}$$

where ε denotes a small parameter. In general $\mathcal{H}_{A}$ will contain fourth-order or even higher-order terms.

Let us assume that one of the second quantization operators appearing in $\mathcal{H}_{(2)}$, say A, satisfies the equation

$$[A, \mathcal{H}_{(2)}] = L_1 A, \tag{4.15}$$

where L_1 designates some number or a linear operator at most. The relevant equation of motion can be written

$$i\frac{dA}{dt} = L_1 A + \varepsilon R_1 A_1, \tag{4.16}$$

where R_1 designates some number or a linear operator at most. The quantity A_1 will contain more second quantization operators than the previous one. Hence using equation (4.13) one has

$$(E - L_1)\langle\langle A \mid B\rangle\rangle = \frac{1}{2\pi}\langle[A, B]\rangle + \varepsilon R_1 \langle\langle A_1 \mid B\rangle\rangle. \tag{4.17}$$

The new Green function appearing on the right side of equation (4.17) satisfies a similar equation of motion,

$$i\frac{dA_1}{dt} = L_2 A_1 + \varepsilon R_2 A_2,$$

$$(E - L_2)\langle\langle A_1 \mid B\rangle\rangle = \frac{1}{2\pi}\langle[A_1, B]\rangle + \varepsilon R_2 \langle\langle A_2 \mid B\rangle\rangle. \tag{4.18}$$

As in the previous case R_2 designates some number or a linear operator at most. Also the new quantity A_2 will contain more second quantization operators than the previous one. In principle the above prescription can be continued up to the required accuracy by cutting-off the equations of motion at the order n,

$$(E-L_{n+1})\,\langle\langle A_n \mid B\rangle\rangle = \frac{1}{2\pi}\langle[A_n, B]\rangle,$$

$$\langle\langle A_{n+1} \mid B\rangle\rangle = 0 \tag{4.19}$$

where $n \geqslant 1$.

The above developed formalism will be used to calculate the energy of elementary excitations by considering the chain for the first two Green functions. Writing

$$G_1 = \langle\langle A \mid B\rangle\rangle, \qquad G_2 = \langle\langle A_1 \mid B\rangle\rangle,$$

$$\langle\langle A_2 \mid B\rangle\rangle = 0 \tag{4.20}$$

equations (4.17) and (4.18) lead to

$$(E-L_1)\,G_1 = I_1+\varepsilon R_1 G_2,$$

$$(E-L_2)\,G_2 = I_2, \qquad G_3 = 0, \tag{4.21}$$

where

$$I_1 = \frac{1}{2\pi}\langle[A, B]\rangle, \qquad I_2 = \frac{1}{2\pi}\langle[A_1, B]\rangle. \tag{4.22}$$

Supposing now that the first-order and second-order Green functions, G_1 and G_2, are known then the equation of motion for the former one is written in a condensed form

$$(E-L_1-M)G_1 = I_1 \tag{4.23}$$

where the introduced quantity M is named the mass operator. Having substituted (4.21) into (4.23) we obtain the quantity M, in zeroth-order approximation,

$$M \cong \varepsilon R_1 G_2^{(0)} G_1^{(0)-1} \tag{4.24}$$

with

$$G_i^{(0)} = (E-L_i)^{-1} I_i, \qquad i = 1, 2.$$

The energy of the system is given by the pole of the Green function; so, in zeroth-order approximation,

$$E \cong L_1 + M. \tag{4.25}$$

This furnishes the formal perturbation expansion for the Green functions.

4.3. Ferromagnetism at low temperatures

The exposed method will be used to calculate the energy of elementary excitations of an ideal cubic ferromagnet at zero temperature with the nearest-neighbor Heisenberg exchange interactions. In most research papers concerning the present subject the spin operators are treated as boson fields (see Holstein and Primakoff, Dyson, and Maleev as cited in the previous chapter) with the exception of spin one-half which is treated as a fermion field (see Kenan[5] and Gosar[6]). In either case Goldstone's theorem applies, see Stern,[7] which is exposed as follows: no energy gap is expected in the dispersion law for the Heisenberg ferromagnet as $\boldsymbol{q}$ tends to zero,

$$\lim_{\boldsymbol{q}\to 0} E(\boldsymbol{q}) = 0. \tag{4.26}$$

In the following calculations we shall consider only the boson fields.

The Dyson–Maleev representation is given by

$$\begin{aligned} S_f^z &= S - n_f, \\ S_f^+ &= \sqrt{(2S)}\, b_f, \\ S_f^- &= \sqrt{(2S)}\, b_f^\dagger (1 - n_f/2S), \\ n_f &= b_f^\dagger b_f. \end{aligned} \tag{4.27}$$

So the Heisenberg ferromagnet for a weak external magnetic field at zero temperature is transformed into

$$\begin{aligned} \mathscr{H} = {} & \mu H \sum_f n_f + 2S \sum_{\langle fg \rangle} J_{fg}(n_f - b_f^\dagger b_g) \\ & - \sum_{\langle fg \rangle} J_{fg}(n_f n_g - \tfrac{1}{2}\, b_f b_g^\dagger n_g - \tfrac{1}{2}\, b_f^\dagger n_f b_g) \end{aligned} \tag{4.28}$$

where a constant term is suppressed. Here every interaction contribution is counted only once as indicated by the symbol $\langle fg \rangle$. An application of equation (4.13) yields

$$EG_1 = \frac{1}{2\pi} \langle [b_f, b_h^\dagger] \rangle + \tilde{G} \tag{4.29}$$

where

$$\begin{aligned}
G_1 &= \langle\langle b_f \mid b_h^\dagger \rangle\rangle, \\
\tilde{G} &= \langle\langle [b_f, \mathscr{H}]\, b_h^\dagger \rangle\rangle = \sum_{j=1}^{4} \tilde{G}_j, \\
\tilde{G}_1 &= (\mu H + 2zSJ)\, G_1, \\
\tilde{G}_2 &= -2S \sum_g J_{fg} \langle\langle b_g \mid b_h^\dagger \rangle\rangle, \\
\tilde{G}_3 &= -\sum_g J_{fg} \langle\langle b_f n_g \mid b_h^\dagger \rangle\rangle, \\
\tilde{G}_4 &= \sum_g J_{fg} \langle\langle n_f b_g \mid b_h^\dagger \rangle\rangle.
\end{aligned} \tag{4.30}$$

Here all sums are extended over the z nearest neighbors sitting on the cubic lattice sites. The former two Green functions appearing in (4.30) can be calculated immediately with the result

$$\begin{aligned}
\tilde{G}_1 &= (\mu H + 2zSJ)\, N^{-1} \sum_{qq'} \langle\langle b_q \mid b_{q'}^\dagger \rangle\rangle \,\mathrm{EXP} \\
\tilde{G}_2 &= -2zSN^{-1} \sum_{qq'} J(\boldsymbol{q})\, \langle\langle b_q \mid b_{q'}^\dagger \rangle\rangle \,\mathrm{EXP}
\end{aligned} \tag{4.31a}$$

where

$$\mathrm{EXP} = \exp\,[i(\boldsymbol{q} \cdot \boldsymbol{R}_f - \boldsymbol{q}' \cdot \boldsymbol{R}_h)], \qquad f \neq h. \tag{4.31b}$$

The latter two Green functions appearing in (4.30) can be calculated with the help of the Wick–Bloch–Dominicis theorem which is originally formulated by Wick[(8)] Bloch and De Dominicis,[(9)] and Tyablikov.[(3)] This theorem can be exposed for the application to the boson fields as follows: a canonical ensemble average taken over four boson operators is given by

$$\langle ABCD \rangle = \langle AB \rangle \langle CD \rangle + \langle AC \rangle \langle BD \rangle + \langle AD \rangle \langle BC \rangle \tag{4.32}$$

with an obvious notation.‡ Now the Green function $\tilde{G}_3$ can be written

$$\tilde{G}_3 = -\sum_g J_{fg} N^{-2} \sum_{qq'} \exp\left[i(\boldsymbol{q}_1 \cdot \boldsymbol{R}_f - \boldsymbol{q}_2 \cdot \boldsymbol{R}_g + \boldsymbol{q}_3 \cdot \boldsymbol{R}_g - \boldsymbol{q}' \cdot \boldsymbol{R}_h)\right]$$
$$\langle\langle b_{q_1} b^\dagger_{q_2} b_{q_3} \mid b^\dagger_{q'} \rangle\rangle .$$

There are two nonvanishing contributions coming from $\tilde{G}_3$, according to equation (4.32),

(1) $\boldsymbol{q}_2 = \boldsymbol{q}_3, \qquad \boldsymbol{q}_1 = \boldsymbol{q},$

$$\langle\langle b_{q_1} b^\dagger_{q_2} b_{q_3} \mid b^\dagger_{q'} \rangle\rangle = \langle b^\dagger_{q_2} b_{q_2} \rangle \langle\langle b_q b^\dagger_{q'} \rangle\rangle ,$$

and

(2) $\boldsymbol{q}_1 = \boldsymbol{q}_2 = \boldsymbol{q} + \boldsymbol{k}, \qquad \boldsymbol{q}_3 = \boldsymbol{q},$

$$\langle\langle b_{q_1} b_{q_2} b_{q_3} \mid b_{q'} \rangle\rangle = \langle b^\dagger_{q+k} b_{q+k} \rangle \langle\langle b_q \mid b^\dagger_{q'} \rangle\rangle .$$

Therefore

$$\tilde{G}_3 = -zN^{-2} \Big\{ \sum_{q_2 q q'} J \langle n_{q_2} \rangle \langle\langle b_q \mid b^\dagger_{q'} \rangle\rangle \, \mathrm{EXP} + \sum_{kqq'} J(\boldsymbol{k}) \langle n_{q+k} \rangle \langle\langle b_q \mid b^\dagger_{q'} \rangle\rangle \, \mathrm{EXP} \Big\} \tag{4.33}$$

A similar calculation holds for the Green function $\tilde{G}_4$,

$$\tilde{G}_4 = \sum_g J_{fg} N^{-2} \sum_{q_j q'} \exp\left[i(-\boldsymbol{q}_1 \cdot \boldsymbol{R}_f + \boldsymbol{q}_2 \cdot \boldsymbol{R}_f + \boldsymbol{q}_3 \cdot \boldsymbol{R}_g - \boldsymbol{q}' \cdot \boldsymbol{R}_h)\right]$$
$$\langle\langle b^\dagger_{q_1} b_{q_2} b_{q_3} \mid b^\dagger_{q'} \rangle\rangle .$$

As in the previous case there are two nonvanishing contributions coming from $\tilde{G}_4$,

(3) $\boldsymbol{q}_1 = \boldsymbol{q}_2, \qquad \boldsymbol{q}_3 = \boldsymbol{q},$

$$\langle\langle b^\dagger_{q_1} b_{q_2} b_{q_3} \mid b^\dagger_{q'} \rangle\rangle = \langle b^\dagger_{q_1} b_{q_1} \rangle \langle\langle b_q \mid b^\dagger_{q'} \rangle\rangle ,$$

and

(4) $\boldsymbol{q}_1 = \boldsymbol{q}_3, \qquad \boldsymbol{q}_2 = \boldsymbol{q},$

$$\langle\langle b^\dagger_{q_1} b_{q_2} b_{q_3} \mid b^\dagger_{q'} \rangle\rangle = \langle b^\dagger_{q_1} b_{q_1} \rangle \langle\langle b_q \mid b^\dagger_{q'} \rangle\rangle .$$

‡ The above equation is an equivalent to a random-phase approximation.

Therefore

$$\tilde{G}_4 = zN^{-2}\Big\{\sum_{q_1qq'} J(\boldsymbol{q}) \langle n_{q_1}\rangle \langle\langle b_q \mid b^\dagger_{q'}\rangle\rangle \,\mathrm{EXP}$$
$$+\sum_{q_1qq'} J(\boldsymbol{q}_1) \langle n_{q_1}\rangle \langle\langle b_q \mid b^\dagger_{q'}\rangle\rangle \,\mathrm{EXP}\Big\}. \tag{4.34}$$

Upon observing equations (4.29)–(4.34) the energy of elementary excitations can be read-off almost immediately with the result

$$E(\boldsymbol{q}) = \mu H + 2zS[J - J(\boldsymbol{q})]$$
$$-zN^{-1}\sum_{p}[J + J(\boldsymbol{q}-\boldsymbol{p}) - J(\boldsymbol{q}) - J(\boldsymbol{p})]\,\langle n_p\rangle \tag{4.35}$$

where the reciprocal lattice vector $\boldsymbol{p}$ is introduced for $\boldsymbol{q}_2$ and $\boldsymbol{q}+\boldsymbol{k}$ in $\tilde{G}_3$, and for $\boldsymbol{q}_1$ in $\tilde{G}_4$. Although the mass operator method is not mentioned explicitly the above result is in agreement with equation (4.24) for $\varepsilon = 1$ where the energy of elementary excitations is expressed by the mass operator. What is more, the above result is in agreement with that in the previous chapter. Therefore one can observe a neat demonstration of the equivalence existing between the Holstein–Primakoff and Dyson–Maleev representations. It is worth noting that the energy $E(\boldsymbol{q})$ satisfies Goldstone's theorem as stated at the beginning of the present section.

4.4. The kinematical interaction

According to Dyson the energy of elementary excitations apart from free spin waves contains dynamical and kinematical interactions which are defined as follows. The dynamical interaction appears as a result of scattering taking place between free spin waves. At zero temperature it is given exactly by the last term in equation (4.35) appearing as the sum extended over the entire reciprocal lattice space.

On the other hand, the kinematical interaction comes from the non-orthogonality of the spin-wave states which is caused by the fact that "more than $2S$ units of reversed spin cannot be attached to the same atom simultaneously". Hence there is a certain statistical obstruction to any dense concentration of spin waves at a given lattice site. Moreover, Dyson proved the existence of a finite energy gap separating the lowest proper or physically acceptable eigenstates from the lowest improper or physically unacceptable eigenstates, the gap being of the order kT_c at least. Since the physically unacceptable states are a purely artificial object they must disappear from all physically observable quantities. To do this Dyson had to remove the kinematical contributions from the free energy by calculating the partition function which then consisted of two major parts, one taken over all physically acceptable states of an ideal model, the other taken over all physically unacceptable states. The latter part of the partition function is smaller than the former one by the factor

$$\exp\left(-\frac{\varepsilon}{kT}\right)$$

where

$$\varepsilon \cong 2kT_c/3. \tag{4.36}$$

The present method in comparison with the one used by Dyson has a certain advantage in the sense that the spontaneous magnetization at $T = 0$ (see Section 3.5) is calculated by a straightforward summation over the occupation number operator in the reciprocal lattice space. (Dyson has calculated the free energy at first and then he has obtained the spontaneous magnetization by differentiating the free energy with respect to an external magnetic field as its magnitude tends to zero.) To see the kinematical effect within the present scope we analyse the Heisenberg ferromagnet with spins one-half as they are situated at the cubic lattice. Clearly two neighbor spins can be either parallel or antiparallel, the total spin quantum number being equal to 1 or 0, respectively. One writes

$$S_1^z = \tfrac{1}{2} - n_1, \qquad S_2^z = \tfrac{1}{2} - n_2,$$

$$\mathcal{H} = -\frac{J}{2}\left[S(S+1) - S_1(S_1+1) - S_2(S_2+1)\right]$$

with $n_1, n_2 = 0$, or 1. The unphysical region is obtained when one of the spins is "frozen" in the ground state, say $S_1^z = \frac{1}{2}$, whereas the other is allowed to take the values

$$S_2^z = \tfrac{1}{2} - n_2$$

with $n_2 \geqslant 2$. The various values of the z-components as introduced so far are illustrated in Table 4.1. Here one may observe that the two physically acceptable states are separated by the energy J, whereas the lowest physically unacceptable state is separated from the lowest physically acceptable one by the energy $3J/2$. The energy of elementary excitations as given by equation (4.35) can be written

$$E(\boldsymbol{q}) = E_0(\boldsymbol{q}) + \mathcal{H}_{\text{dyn}} \tag{4.37}$$

where the two terms represent the energy of free spin waves and the effect coming from a dynamical interaction, respectively. The lower and upper bounds within the first Brillouin zone can be determined

TABLE 4.1.

Various Values for the Quantities Associated with Two Neighbor Spins in Physical and Unphysical Regions

Spin arrangement	S_1^z	S_2^z	S	$n \equiv n_2$	$\mathcal{H}$ (in J)
↑ ↑	$\frac{1}{2}$	$\frac{1}{2}$	1	0	$-\frac{1}{4}$
↑ ↓	$\frac{1}{2}$	$-\frac{1}{2}$	0	1	$\frac{3}{4}$
↑ ↓↓	$\frac{1}{2}$	$-\frac{3}{2}$	1	2	$\frac{5}{4}$
↑ ↓↓…↓	$\frac{1}{2}$	$-(2n-1)/2$	$(n-1)$	$\geqq 2$	$(2n+1)/2$

as follows:

$$-\pi/a \leqslant q_x, q_y, q_z \leqslant \pi/a$$

$$0 \leqslant E_0(\boldsymbol{q}) \leqslant 2zJ$$

$$(-4zJ/N)\sum_p \langle n_p \rangle \leqslant \mathscr{H}_{\text{dyn}} \leqslant (2\,zZJ/N)\sum_p \langle n_p \rangle \tag{4.38}$$

the sum being extended over the introduced Brillouin zone. Now the largest contribution to the canonical ensemble average

$$\langle n_p \rangle = \left\{ \exp \frac{E(\boldsymbol{p})}{kT} - 1 \right\}^{-1}$$

comes from the long-wavelength spin waves for which $\boldsymbol{p} \to 0$ and $E_0(\boldsymbol{p}) \propto p^2$. However, these terms are separated from the unphysical region by the energy $3J/2$ at least, so the kinematical effect vanishes at zero temperature exponentially, the vanishing factor being

$$\exp(-3J/2kT).$$

An explicit account of the kinematical interaction is taken through the partition function. Writing the total partition function as

$$Z = Z_T - Z_I \tag{4.39}$$

where two terms on the right refer to the Dyson ideal model and an ensemble of improper states, respectively, we obtain

$$\langle S_j^z \rangle = Z_T^{-1}\,\mathrm{Tr}\,S_j^z \exp(-\beta\mathscr{H}) \tag{4.40a}$$

where

$$Z_T^{-1} = (Z+Z_I)^{-1} \cong Z^{-1}(1-Z_I/Z). \tag{4.40b}$$

The above expansion is correct up to a first-order approximation at least. Therefore the thermal average of a given z component of the lattice pseudo-spin to be compared to the observed magnetization relative to a single atomic site is expressed by

$$\langle S_j^z \rangle_T = \langle S_j^z \rangle (1-Z_I/Z) \tag{4.41a}$$

where

$$\frac{Z_I}{Z} = \exp\left(-\frac{3J}{2kT}\right). \tag{4.41b}$$

The above result is correct up to a first-order approximation at least.

The actual comparison of the calculated and observed magnetizations relative to a single atomic site depends somewhat on the exact relationship that is supposed to exist between the exchange integral, J, and the transition temperature, T_c. Here the relevant and hypothetical relationship is established by an extension and repeated use of the equations of motion written for the chain of coupled Green functions. There are at the moment two considerably well-defined and justified decoupling procedures (techniques) whose successive application covers the whole temperature axis, thus starting from the low temperatures and extending over to the critical region as well as the paramagnetic region.

One procedure is due to Tyablikov[3] (see also the work of Tahir-Kheli and Ter Haar cited in the previous chapter), the other one is due to Callen.[10] A vague idea of a distinction between the two computing techniques is gained as follows. On the former decoupling procedure one writes

$$\langle\langle S_g^z S_f^+ \mid B\rangle\rangle = \langle S_g^z\rangle \langle\langle S_f^+ \mid B\rangle\rangle, \tag{4.42}$$

whereas on the latter decoupling procedure one writes

$$\langle\langle S_g^z S_f^+ \mid B\rangle\rangle = \langle S_g^z\rangle \langle\langle S_f^+ \mid B\rangle\rangle - \alpha\langle S_g^- S_f^+\rangle \langle\langle S_g^+ \mid B\rangle\rangle \tag{4.43a}$$

where the decoupling parameter, α, is selected to be

$$\alpha = \frac{\langle S_g^z\rangle}{S}, \qquad \text{spin one-half,}$$

$$\alpha = \frac{1}{2S}\frac{\langle S_g^z\rangle}{S}, \qquad \text{arbitrary spin.} \tag{4.43b}$$

The two computing procedures are manifested most clearly on the problem of the transition temperature. An overall conclusion is reached along the numerical analysis: the transition temperature with Tyablikov's decoupling procedure is considerably different from that obtained with a molecular-field approximation. On the other hand, the transition temperature with Callen's decoupling procedure is closer to that obtained with a molecular-field approximation, in particular for small values of the pseudo-spin and for lattices having a greater number of neighbor atoms, see Table 4.2. This situation is further reflected on the

TABLE 4.2.

Transition Temperature on a Molecular-field Approximation, T_c (MFA), *with Respect to Exchange Coupling,* J, *and Transition Temperature in Various Theories One Relative to Another*

kT_c (MFA)/J

Pseudo-spin	Cubic P	Cubic I	Cubic F
$\frac{1}{2}$	3	4	6
1	8	10.67	16
$\frac{3}{2}$	15	20	30
2	24	32	48
$\frac{5}{2}$	35	46.67	70
3	48	64	96
	T_c (GFAT)/T_c (MFA)		
Arbitrary	0.66	0.72	0.74
	T_c (GFAC)/T_c (MFA)		
$\frac{1}{2}$	0.90	0.92	0.93
1	0.81	0.85	0.87
$\frac{3}{2}$	0.78	0.83	0.85
2	0.77	0.82	0.83
$\frac{5}{2}$	0.766	0.812	0.833
3	0.758	0.806	0.828

Note. The following abbreviations are used: GFAT for "Green-function approximation with Tyablikov's decoupling"; GFAC for "Green-function approximation with Callen's decoupling".

TABLE 4.3.

Observed Relative Room-temperature Magnetization Versus Various Theories for Ferromagnetic Elements; all Contributions Expressed in Units 10^{-2}

Substance	Pseudo-spin	Structure	Observed at 20 °C	GFAT‡ (T = 20 °C)			GFAC§ (T = 20 °C)		
				D‖	K ¶	Total	D‖	K ¶	Total
Fe	1	Cubic I		3.64	−1.82	1.82	4.79	−2.66	2.13
	1	Cubic F	1.76	3.58	−2.28	1.30	4.73	−3.22	1.51
Co	1	Cubic F		2.20	−1.20	1.00	2.87	−1.72	1.15
	$\frac{1}{2}$	Cubic F	0.92	2.81	−0.56	2.25	3.99	−1.09	2.90
Ni	$\frac{1}{2}$	Cubic F	5.41	10.99	−5.30	5.69	14.62	−8.20	6.42

‡ Green-function approximation with Tyablikov's decoupling. § Green-function approximation with Callen's decoupling, ‖ Dynamical contribution, ¶ Kinematical contribution.

kinematical contribution to the magnetization as related to a single atomic site. A numerical analysis is carried out to show the relative room-temperature magnetization for well-known ferromagnetic elements (iron, cobalt, nickel). It turns out that the magnetization is closer to observed values if computed with Tyablikov's decoupling procedure rather than with that of Callen (see Table 4.3).

4.5. Antiferromagnetism, I

As implied by symmetry arguments the magnetization of an ideal antiferromagnet at zero temperature vanishes identically since the spins on one sublattice are antiparallel exactly to those on the other sublattice. However, this does not mean that all the spins on each sublattice must necessarily point to the same direction with respect to an external magnetic field. As demonstrated by Marshall(11) in two exhaustive papers using a number of convincing approximations certain antiferromagnetic structures (notably the linear chain, simple plane, simple cubic, as well as body-centered cubic) have every one of them a disordered ground state. Therefore the spins on a given sublattice do not necessarily point in one and the same direction making antiferromagnetism apparently impossible. One can readily apply the Green functions method in order to see just to what extent the preceding statement is true.

For this reason the hamiltonian of Section 3.7 with nearest-neighbor interactions is taken over in the form

$$\begin{aligned}
\mathcal{H} &= \mathcal{H}_{(2)}+\mathcal{H}_{(4)} \\
\mathcal{H}_{(2)} &= \sum_{\boldsymbol{q}} [\lambda(n_{\boldsymbol{q}}+m_{\boldsymbol{q}})+\mu(\boldsymbol{q})(\xi_{\boldsymbol{q}}+\eta_{\boldsymbol{q}})] \\
\mathcal{H}_{(4)} &= N^{-1}\sum_{\boldsymbol{q}\boldsymbol{p}} \{\tfrac{1}{2}[J(\boldsymbol{q})\, c_{\boldsymbol{q}}^{\dagger}c_{\boldsymbol{q}}c_{\boldsymbol{p}}d_{\boldsymbol{p}}+J(\boldsymbol{q})\, c_{\boldsymbol{q}}d_{\boldsymbol{p}}^{\dagger}d_{\boldsymbol{p}}d_{\boldsymbol{q}}]+Jc_{\boldsymbol{q}}^{\dagger}c_{\boldsymbol{q}}d_{\boldsymbol{p}}^{\dagger}d_{\boldsymbol{p}} \\
&\quad +J(\boldsymbol{q}-\boldsymbol{p})\, c_{\boldsymbol{q}}^{\dagger}c_{\boldsymbol{p}}d_{\boldsymbol{q}}^{\dagger}d_{\boldsymbol{p}}\} \qquad (4.44)
\end{aligned}$$

$$\lambda = zSJ, \quad \mu(q) = zSJ(q), \quad J_{AA} = J_{BB} = 0,$$
$$J_{AB} = J > 0, \quad n_q = c_q^\dagger c_q, \quad m_q = d_q^\dagger d_q, \quad \xi_q = c_q^\dagger d_q^\dagger,$$
$$\eta_q = c_q d_q, \tag{4.45}$$

where all the introduced operators are represented by boson fields. (Here N designates the total number of sites on each sublattice.) Let us introduce the following Green function matrix:

$$\hat{G}(\tau) = \begin{bmatrix} G_{11}(\tau) & G_{12}(\tau) \\ G_{21}(\tau) & G_{22}(\tau) \end{bmatrix} \tag{4.46}$$

where

$$\begin{aligned} G_{11}(\tau) &= \langle Tc_q(\tau)\, c_q^\dagger(0)\rangle, \\ G_{12}(\tau) &= \langle Tc_q(\tau)\, d_q(0)\rangle, \\ G_{21}(\tau) &= \langle Td_q^\dagger(\tau)\, c_q^\dagger(0)\rangle, \\ G_{22}(\tau) &= \langle Td_q^\dagger(\tau)\, d_q(0)\rangle. \end{aligned} \tag{4.47}$$

Here T is the Wick time-ordering operator, τ is an imaginary time in units $\hbar = 1$, $\tau = it$. To simplify further computations the usual factor $-i$ appearing as part of the causal double-time temperature-dependent Green function will be suppressed.

According to the method of Abrikosov *et al.*[12] the Fourier transform of a given Green function component, $G_{\mu\nu}(\tau)$ $(\mu, \nu = 1, 2)$, can be written

$$G_{\mu\nu}(\tau) = \frac{1}{\beta} \sum_{n=-\infty}^{\infty} g_{\mu\nu}(n) \exp(-iE_n\tau) \tag{4.48}$$

where

$$g_{\mu\nu}(n) = \int_0^\beta G_{\mu\nu}(\tau) \exp(iE_n\tau)\, d\tau,$$

$$E_n = \frac{2\pi}{\beta} n = 2\pi nkT. \tag{4.49}$$

If τ tends to zero equation (4.48) is reduced to

$$G_{\mu\nu}(0) = \frac{1}{\beta} \sum_{n=-\infty}^{\infty} g_{\mu\nu}(n). \tag{4.50}$$

It is easy to show that the following relationship holds

$$G_{\mu\nu}(\tau) = G_{\mu\nu}(-\tau)$$

which applies to the boson fields. Also

$$\begin{aligned} g_{11}(n) &= g_{22}(-n), \\ g_{12}^{*}(n) &= g_{21}(n). \end{aligned} \tag{4.51}$$

In general the following equations of motion hold

$$\frac{d\hat{G}(\tau)}{d\tau} = \pm\delta(\tau) + \langle T[\mathscr{H}, A(\tau)]\, B(0)\rangle$$

if

$$[A, B] = \pm 1,$$

and

$$\frac{d\hat{G}(\tau)}{d\tau} = \langle T[\mathscr{H}, A(\tau)]\, B(0)\rangle$$

if

$$[A, B] = 0$$

where A and B designate arbitrary spin-wave operators. Therefore the Green function matrix satisfies the equation of motion as follows:

$$\frac{d\hat{G}(\tau)}{d\tau} = \begin{bmatrix} 1 & 0 \\ 0 & -1 \end{bmatrix} \delta(\tau) + \begin{bmatrix} \left\langle T \dfrac{dc_p(\tau)}{d\tau}\, c_p^{\dagger}(0) \right\rangle; & \left\langle T \dfrac{dc_p(\tau)}{d\tau}\, d_p(0) \right\rangle \\ \left\langle T \dfrac{dd_p^{\dagger}(\tau)}{d\tau}\, c_p^{\dagger}(0) \right\rangle; & \left\langle T \dfrac{dd_p^{\dagger}(\tau)}{d\tau}\, d_p(0) \right\rangle \end{bmatrix} \tag{4.52}$$

where

$$\frac{dA(\tau)}{d\tau} = [\mathscr{H}, A(\tau)].$$

As proved by Anderson(13) (who considered an approximate quantum theory of the antiferromagnetic ground state in detail) the sublattice spin in the ground state does not maintain a constant direction of motion, the indeterminacy in direction being in agreement with our knowledge of singlet states. To represent the state of a long-range order on the entire lattice one has to calculate the total spin on a given sub-lattice. (Clearly an average spin par atom parallel to the total sub-lattice spin can be measured by a neutron diffraction method.) To see how much the antiferromagnetic ground state is actually disordered we introduce the relative sublattice order parameter,

$$\sigma = \frac{\langle S^z_{(1)}\rangle - \langle S^z_{(2)}\rangle}{2SN} \tag{4.53}$$

where the total z-component of the spin of a sublattice is denoted by $S^z_{(1)}$ [or $S^z_{(2)}$ for the second sublattice]. By virtue of equations (4.45)–(4.47) the above expression can be written

$$\sigma = 1-(2SN)^{-1}\sum_p [\langle Tc_p(\tau)\, c_p^\dagger(0)\rangle + \langle Td_p^\dagger(\tau)\, d_p(0)\rangle - 1] \tag{4.54}$$

as τ tends to zero. Clearly the maximum value of σ is equal to one.

To complete (4.54) in a numerical form we will neglect the hamiltonian $\mathscr{H}_{(4)}$, equation (4.44), on the grounds that an application of the Wick–Bloch–Dominicis statistical theorem will produce second-order terms in the low-temperature region at least. Using equations (4.48) and (4.52) we arrive at the following relation:

$$\frac{d\hat{G}(\tau)}{d\tau} = \begin{bmatrix} 1 & 0 \\ 0 & -1 \end{bmatrix} \delta(\tau) - \lambda \begin{bmatrix} 1 & \gamma(\boldsymbol{p}) \\ -\gamma(\boldsymbol{p}) & -1 \end{bmatrix} \hat{G}(\tau) \tag{4.55}$$

where

$$\gamma(\boldsymbol{p}) = \mu(\boldsymbol{p})/\lambda = J(\boldsymbol{p})/J,$$

$\boldsymbol{p}$ being a reciprocal lattice vector. The solution to this equation in terms of the Fournier transform is given by

$$\begin{bmatrix} g_{11}(n) & g_{12}(n) \\ g_{21}(n) & g_{22}(n) \end{bmatrix} = \frac{1}{E_n^2+E^2(\boldsymbol{p})} \begin{bmatrix} \lambda+iE_n & -\lambda\gamma(\boldsymbol{p}) \\ -\lambda\gamma(\boldsymbol{p}) & \lambda-iE_n \end{bmatrix}$$

where

$$E(\boldsymbol{p}) = zSJ\sqrt{\{1-\gamma^2(\boldsymbol{p})\}}. \tag{4.56}$$

By definition

$$\sigma = 1-(2SN)^{-1}\sum_{p}\sum_{n=-\infty}^{\infty}\left[\frac{g_{11}(n)+g_{22}(n)}{\beta}-1\right]$$

or

$$\sigma = 1-\frac{c}{2S} \tag{4.57}$$

with‡

$$c = \frac{\lambda}{2\pi kTN}\sum_{p}\left(\frac{\coth \pi x}{x}\right)-1, \tag{4.58a}$$

$$x = \frac{\lambda\sqrt{\{1-\gamma^2(\boldsymbol{p})\}}}{2\pi kT}. \tag{4.58b}$$

By a low-temperature expansion as T tends to zero the sublattice order parameter is reduced to the numerical computation of an integral,

$$c = N^{-1}\sum_{p}[1-\gamma^2(\boldsymbol{p})]^{-1/2}-1. \tag{4.59}$$

A method used to compute the quantity c for body-centered or simple cubic lattices is exposed in Appendix D, the accuracy being one part in 10^5, with the result

$$c = 0.11863 \quad \text{or} \quad 0.15619, \tag{4.60}$$

respectively.

The zero-temperature sublattice order parameter depends on the lattice structure under consideration as well as on the magnitude of

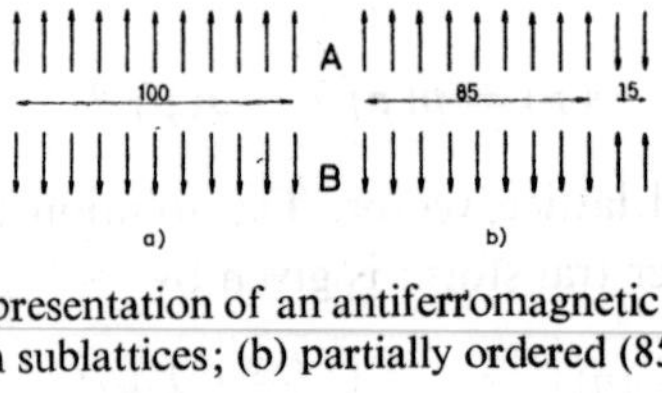

FIG. 4.1. Schematic representation of an antiferromagnetic ground state: (a) completely ordered on both sublattices; (b) partially ordered (85%) on each sublattice.

‡ A use is made of the following equation:

$$\sum_{n=-\infty}^{\infty}\frac{1}{n^2+x^2} = \frac{\pi\coth \pi x}{x}, \qquad x > 0.$$

the pseudo-spin. As an illustration we quote below the first three values of $\sigma(0)$ on the face-centered cubic lattice with a simple cubic configuration (see Appendix D and Fig. 4.1):

$$\sigma(0) = \begin{cases} 0.85 & \text{if} \quad S = \frac{1}{2}, \\ 0.92 & \text{if} \quad S = 1, \\ 0.95 & \text{if} \quad S = \frac{3}{2}. \end{cases} \tag{4.61}$$

Therefore the zero-temperature sublattice magnetization, $M_s(0)$, is never equal exactly to the sublattice magnetization for an ideally ordered system, M_∞. The latter quantity is defined by

$$M_\infty = Ng\mu_B S \tag{4.62}$$

where the used symbols have their usual meaning. A general numerical result which relates the two introduced sublattice magnetizations is presented in Table 4.4 as obtained by various methods.

TABLE 4.4.

Sublattice Magnetization at Zero Temperature Obtained by Various Methods

	NaCl structure	CsCl structure
$\dfrac{M_s(0)}{M_\infty/S} =$	$S-0.078$‡, §, ‖ $S-0.078(1)$ ¶	$S-0.075$§, ‖ $S-0.059(3)$ ¶

‡ After Anderson.[13] § After Kubo,[14] ‖ After Oguchi, cited in Chapter 3, ¶ Present method, see Appendix D.

4.6. Antiferromagnetism, II

The presented results on the zero-temperature sublattice magnetization are in agreement with those of Liu and Oguchi (both are cited in Chapter 3) as well as with those of Anderson[13] and Kubo.[14]

However, a general expansion replacing equation (4.57) for the whole low-temperature region with an exact account of the neglected hamiltonian would bring the sublattice order parameter to the form

$$\sigma(T) = \sigma(0) - \sigma_1\tau^2 - \sigma_2\tau^6 + \dots \tag{4.63}$$

where τ designates a dimensionless temperature,

$$\tau = \frac{kT}{2SJ}.$$

Using the iteration method as developed in the cited references one is able to estimate the expansion coefficients σ_1 and σ_2, an order of magnitude at least. It requires a laborious work to do so, therefore we shall content ourselves to quote the result for a body-centered cubic configuration (CsCl structure),

$$\sigma_1 \cong 3 \times 10^{-2},$$

whereas

$$\sigma_2 \cong 10^{-4}.$$

It is worth noting that an additional expansion term of the form τ^4 is possible to obtain on a somewhat different iteration basis, see Dembinski.[15]

In conclusion, the present spin-wave theory predicts the sublattice order parameter, σ, and therefore the sublattice magnetization in good agreement with the perturbation theory of Marshall.[11] What is more Thouless[16] proved that the "paramagnetic" ground-state wave functions proposed by Marshall in the cited papers for a system of spin one-half particles interacting through an antiferromagnetic Heisenberg exchange coupling produce a large sublattice magnetization, comparable to the magnetization as predicted by the spin-wave theory and yet very close to the result given by the perturbation theory. In the cited paper Thouless has pointed out that the problem of a long-range order in an antiferromagnetic system is closely related to the problem of a long-range order in a system of interacting bosons.

In order to compute the low-temperature expansion for the order parameter and specific heat we recall a well-known analogy that exists between antiferromagnetism and lattice dynamics (see Table 4.5). Writing the order parameter

$$\sigma(T) = \left(1-\frac{1}{2S}\right) - (2SN)^{-1}\sum_{p} \coth(\pi x)[1-\gamma^2(\boldsymbol{p})]^{-1/2} \quad (4.64)$$

TABLE 4.5.

Antiferromagnetism Versus Lattice Dynamics

Physical quantity	Lattice dynamics	Antiferromagnetism
Elementary excitations	Acoustic phonons	Antiferromagnetic magnons
Momentum	$\hbar \boldsymbol{p}$	$\hbar \boldsymbol{p}$
Energy	$\hbar c p$	$\hbar c p$
First characteristic quantity	Velocity of elastic waves; c	Velocity of spin waves; c
Order of magnitude	10^5 cm/sec	10^5 cm/sec
Second characteristic quantity	Debye temperature; Θ_D	Néel temperature; Θ_N
Expression	$c\hbar/ka$	$c\hbar/ka$
Order of magnitude	500 K	500 K

and using the formal identity

$$\coth(\pi x) = 1+\frac{2}{\exp(2\pi x)-1}$$

we arrive at the expression

$$\sigma(0)-\sigma(T) = (SN)^{-1}\sum_{p}\langle n_p\rangle[1-\gamma^2(\boldsymbol{p})]^{-1/2} \quad (4.65a)$$

where

$$\langle n_p \rangle = \frac{1}{\exp(2\pi x) - 1} \tag{4.65b}$$

A low-temperature approximation is obtained if only small values of $\boldsymbol{p}$ are taken into account, thus leading to

$$\langle n_p \rangle = \frac{1}{\exp \dfrac{E(\boldsymbol{p})}{kT} - 1} \tag{4.66a}$$

where

$$E(\boldsymbol{p}) = zSJ\,|\,\boldsymbol{p}\,|\,\sqrt{\left(\frac{2}{z}\right)}\,a. \tag{4.66b}$$

Naturally the quantity $E(\boldsymbol{p})$ is interpreted as the energy of normal antiferromagnetic modes ("antiferromagnetic" spin waves) whose momentum is equal to $\hbar p$. The spin-wave traveling velocity is related to the energy of normal antiferromagnetic modes in the same way as the elastic-wave traveling velocity is related to the energy of acoustic phonons.

One can postulate

$$E(\boldsymbol{p}) = k\Theta_N \frac{p}{p_{max}} \tag{4.67}$$

with Θ_N being a quantity of the order of magnitude of the Néel temperature, whereas p_{max} depends on the lattice structure under consideration. For a face-centered cubic lattice with a simple cubic configuration (NaCl structure) we may write

$$p_{max} = \pi/a,$$

a being the lattice constant. Now the total number of atoms on a given sublattice per unit volume is

$$n = (2\pi)^{-3}\, 4\pi(p_{max})^3/3 = (p_{max})^3/(6\pi^2). \tag{4.68}$$

Therefore the low-temperature expansion for the sublattice order parameter can be written

$$\sigma(0)-\sigma(T) \cong \frac{\sqrt{3}}{2\pi^2 aS}\frac{V}{N}\int_0^\infty \frac{p\,dp}{\exp\dfrac{\Theta_N p}{Tp_{\max}}-1}$$

$$= \frac{\sqrt{3}}{2S}\frac{V}{Na^3}\Gamma(2)\,\zeta(2)\left(\frac{T}{\Theta_N}\right)^2. \tag{4.69}$$

Here the used symbols have their usual meaning.

Also the specific heat can be calculated in a straightforward manner. By definition the internal energy of an antiferromagnetic system can be written (by invoking the present analogy established between anti-ferromagnetism and lattice dynamics) as follows:

$$\langle \mathcal{H} \rangle = N^{-1}\sum_{p} E(\boldsymbol{p})\langle n_{\boldsymbol{p}}\rangle.$$

The specific heat is then expressed as a derivation with respect to temperature,

$$C(T) = \frac{\partial \langle \mathcal{H} \rangle}{\partial T}.$$

The result is given by, again for a face-centered cubic lattice with a simple cubic configuration (NaCl structure),

$$C(T) = \frac{2\pi}{\Theta_N}\frac{V}{Na^3}\Gamma(4)\,\zeta(4)\left(\frac{T}{\Theta_N}\right)^3. \tag{4.70}$$

This expansion holds for all temperatures for which

$$T \ll \Theta_N.$$

4.7. Experimental evidence

There are two large categories of experimental evidence, according to the review article by Nauciel–Bloch,[17] one having the intention to test the validity range of the theory, the other having the intention to investigate new magnetic materials. The former category is usually related to insulators with simple structures the cations of which are in a definite spin state and the dominant interactions are described by a Heisenberg exchange model with a small number of exchange integrals. A comparison between the experimental and theoretical results on temperature-dependent magnetic quantities is possible only for a very limited number of ferromagnets (EuO, EuS) or antiferromagnets (MnF_2). The latter category is concerned with the interpretation of the spin-wave theory in relation to the other experimental data as obtained at low and medium temperatures in order to test directly the fundamental assumptions of the exchange integrals. Except for very few substances (e.g. MnF_2) there is no sufficient knowledge of the nature and magnitude of the exchange interaction beyond the usual spin-wave theory. Moreover, most of the magnetic materials do not possess simple structures to allow for a precise calculation without a specific model. Therefore an average interpretation of the experimental evidence is inclined to establish an internal consistency of the theory rather than to notify an absolute change. The best test of the theory is the comparison of the temperature dependent spin-wave excitation energies. In this sense an inelastic neutron measurement was performed on MnF_2 to demonstrate an excellent agreement between the theory and experiment. However, very often the neutron scattering techniques cannot be applied if the magnetic target contains an element with a large cross section. In this case a new technique must be used (e.g. nuclear magnetic resonance).

Both theoretical and experimental results on the antiferromagnetic resonance frequency, sublattice magnetization, and magnetic specific heat of FeF_2 and MnF_2 are presented by Nagai and Tanaka.[18]

Experimental data are mostly found somewhere between two temperature-dependent magnon-energy theories, one being the conventional random-phase approximation, the other being the magnon-renormalization approximation. A calculation of the magnon energy at the first Brillouin zone edge in the case of MnF_2 seems to support the view that the magnon interaction effect is overestimated by the latter method.

A clear evidence for a comparatively large zero-point spin deviation is produced by Rubinstein and Folen[19] who measured a nuclear magnetic resonance in the two-dimensional antiferromagnet, K_2MnF_4, at 4.2 K, and an electron spin resonance of the transferred hyperfine interactions in K_2ZnF_4: $Mn^{2(+)}$. A further evidence is revealed by Loopstra *et al.*[20] who used a neutron diffraction method to determine the magnetic moment of the magnetic ions in K_2MnF_4 at the same temperature. The observed moment is 4.54 instead of 5 Bohr magnetons. This seems to agree with the spin-wave theory.

4.8. The ferroelectric modes

In the present section we will establish under what conditions the soft mode on the Ising model with a transverse field (also named the "tunneling" model) can exist. Assuming nearest-neighbor interactions on a cubic lattice we write the hamiltonian

$$-\mathcal{H} = 2\Omega \sum_f S_f^x + \sum_{\langle fg \rangle} J_{fg} S_f^z S_g^z \tag{4.71}$$

where 2Ω and J_{fg} designate the well-known kinetic and potential energies. The crucial step here is to apply the following two transformations successively, both in the pseudo-spin space; one being a rotation of the pseudo-spin coordinates to the new ones, the other being a suitable representation of the new pseudo-spin coordinates by the boson fields of the Dyson–Maleev type. Therefore

$$S_f^x = \cos\phi S_f^{x'} + \sin\phi S_f^{z'},$$
$$S_f^z = -\sin\phi S_f^{x'} + \cos\phi S_f^{z'}, \tag{4.72}$$

where

$$S_f^{z'} = S - n_f,$$
$$S_f^{x'} = \frac{\sqrt{2S}}{2}\left[b_f + b_f^\dagger\left(1 - \frac{n_f}{2S}\right)\right],$$
$$n_f = b_f^\dagger b_f. \tag{4.73}$$

Here the rotation angle ϕ is a temperature-dependent but a lattice site-independent quantity. An application to the hydrogen-bonded ferroelectrics requires inserting $S = \frac{1}{2}$ throughout the preceding equations but the present remark is correct regardless of an actual magnitude of the spin.

The first transformation gives

$$-\mathscr{H} = 2\Omega\cos\phi\sum_f S_f^{x'} - \sin\phi\cos\phi\sum_{\langle fg\rangle} J_{fg}(S_f^{x'}S_g^{z'}$$
$$+ S_f^{z'}S_g^{x'}) + 2\Omega\sin\phi\sum_f S_f^{z'} + \sum_{\langle fg\rangle} J_{fg}(\sin^2\phi S_f^{x'}S_g^{x'}$$
$$+ \cos^2\phi S_f^{z'}S_g^{z'}). \tag{4.74}$$

In order that the ground state of the pseudo-spin system exists we demand for the first two terms of the hamiltonian (4.74) to vanish identically. There follows the condition:

$$\Omega = \zeta SJ\sin\phi \tag{4.75}$$

with

$$S_f^{z'} = S_g^z = S$$

being substituted in the second sum of (4.74). Here ζ is the number of nearest neighbors. So the hamiltonian (4.74) owing to the condition (4.75) is transformed into

$$-\mathscr{H} = 2\zeta SJ\sin^2\phi\sum_f S_f^{z'} + \sum_{\langle fg\rangle} J_{fg}(\sin^2\phi S_f^{x'}S_g^x$$
$$+ \cos^2\phi S_f^{z'}S_g^{z'}). \tag{4.76}$$

It is easy to see that the kinematical effect has a minor importance owing to a large energy gap separating the physically acceptable states from those physically unacceptable. Indeed, let us consider two spin one-half neighbors,

$$-\mathcal{H} = [(S_1^{z'}+S_2^{z'})\sin^2\phi+\tfrac{1}{2}(\boldsymbol{S}_1\cdot\boldsymbol{S}_2-S_1^{z'}S_2^{z'})\sin^2\phi + S_1^{z'}S_2^{z'}\cos^2\phi]\,J, \tag{4.77}$$

J being an effective exchange coupling. Matrix elements of $\mathcal{H}$ in terms of J according to equation (4.77) for the lowest physical as well as unphysical states are given in Table 4.6.

TABLE 4.6.

Matrix Elements of $\mathcal{H}$ for Two Spin One-half Neighbors, Equation (4.77)

$S_1^{z'}$	$S_2^{z'}$	$\mathcal{H}$ (in J)	Difference (in J)
$\frac{1}{2}$	$\frac{1}{2}$	$-(1+3\sin^2\phi)/4$	0
$\frac{1}{2}$	$-\frac{1}{2}$	$\frac{1}{4}$	$(2+3\sin^2\phi)/4$
$\frac{1}{2}$	$-\frac{3}{2}$	$(3+2\sin^2\phi)/4$	$(4+5\sin^2\phi)/4$
$\frac{1}{2}$	$-\frac{5}{2}$	$(5+4\sin^2\phi)/4$	$(6+7\sin^2\phi)/4$

Looking at Table 4.6 one can observe two important properties:

1. the first two physically acceptable states are separated by the energy $J/2$ at least ($\sin\phi = 0$), or $5J/4$ at most ($\sin\phi = 1$);
2. the lowest physically unacceptable state is separated from the highest physically acceptable one by the energy $J/2$ at least ($\sin\phi = 0$), or J at most ($\sin\phi = 1$).

In both cases the upper energy limit as obtained for $\sin\phi = 1$ ensures immediately the ground-state existence condition, equation (4.75). In conclusion, the kinematical effect is very small in a low-temperature region just as in the case of ferromagnetism.

The second transformation as given by equations (4.73) leads to

$$\mathcal{H} = E_0+\mathcal{H}_{(2)}$$

where

$$E_0 = -\zeta S^2 NJ \cos^2 \phi,$$
$$\mathscr{H}_{(2)} = 2\zeta SJ \sum_f b_f^\dagger b_f - \tfrac{1}{2} S \sum_{\langle fg \rangle} J_{fg}(b_f^\dagger b_g + b_f^\dagger b_g + b_f b_g^\dagger + b_f b_g) \sin^2 \phi. \tag{4.78}$$

Higher-order terms are neglected owing to the conclusion following Table 4.6. Let us introduce the temperature-dependent first-order Green function, as in Section 4.3,

$$G = \langle\langle b_f \mid b_h^\dagger \rangle\rangle,$$
$$\tilde{G} = \langle\langle b_f^\dagger \mid b_h^\dagger \rangle\rangle, \tag{4.79}$$

which satisfy the equations of motion

$$EG = \frac{1}{2\pi} \Delta(f-g) + 2\zeta SJG - S \sum_g J_{fg}(\langle\langle b_g \mid b_h^\dagger \rangle\rangle + \langle\langle b_g^\dagger \mid b_h^\dagger \rangle\rangle) \sin^2 \phi,$$
$$E\tilde{G} = -2\zeta SJ\tilde{G} + S \sum_g J_{fg}(\langle\langle b_g^\dagger \mid b_h^\dagger \rangle\rangle + \langle\langle b_g \mid b_h^\dagger \sin^2 \rangle\rangle)\phi. \tag{4.80}$$

For the lattice sites $h \neq f$ the Fourier transform leads to

$$EG = 2\zeta SJG - \zeta SJ(\boldsymbol{q})\,(G+\tilde{G}) \sin^2 \phi,$$
$$E\tilde{G} = -2\zeta SJ\tilde{G} + \zeta SJ(\boldsymbol{q})\,(\tilde{G}+G) \sin^2 \phi. \tag{4.81}$$

The energy of elementary excitations is thus determined by

$$\begin{vmatrix} E-\Delta(\boldsymbol{q}) & V(\boldsymbol{q}) \\ -V(\boldsymbol{q}) & E+\Delta(\boldsymbol{q}) \end{vmatrix} = 0, \tag{4.82}$$

where

$$\Delta(\boldsymbol{q}) = 2\zeta SJ - SJ(\boldsymbol{q}) \sin^2 \phi,$$
$$V(\boldsymbol{q}) = \zeta SJ(\boldsymbol{q}) \sin^2 \phi.$$

The solution to the above determinant becomes

$$\begin{aligned} E^2(\boldsymbol{q}) &= \Delta^2(\boldsymbol{q}) - V^2(\boldsymbol{q}) \\ &= (2\zeta SJ)^2\{1-\gamma(\boldsymbol{q})\sin^2\phi\} \\ \gamma(\boldsymbol{q}) &= J(\boldsymbol{q})/J(0). \end{aligned} \tag{4.83}$$

Therefore every "tunneling" model when applied to the crystals of cubic symmetry with at least one type of atoms which possess two or more equilibrium positions contains the elementary excitations whose energy vanishes as $\boldsymbol{q} \to 0$, and $\sin\phi \to 1$. This type of motion is also named the soft mode. To the best of our knowledge it is associated not only with crystals which undergo the ferroelectric phase transition (e.g. hydrogen-bonded ferroelectrics) but also with those which undergo the structural phase transition (e.g. strontium titanate[21]). In either case if a certain type of atoms is described by the pseudo-spin method then the total pseudo-spin hamiltonian by virtue of general equations of motion can be diagonalized to obtain a soft-mode energy. As temperature rises up from the transition point the quantity $\sin^2\phi$ acquires the characteristic temperature variation owing to the relative lattice volume expansion. So we write

$$E^2(\boldsymbol{q}) = (2\zeta SJ)^2\left\{\lambda(T-T_c)+\frac{(qa)^2}{\zeta}+\dots\right\} \tag{4.84}$$

λ being the expansion coefficient with an order of magnitude 10^{-4}/degree. (As for the expansion $\lambda(\boldsymbol{q})$ in terms of the reciprocal lattice vector, see Appendix A.)

It is worth noting that the intensity of the coherent nearly elastic neutron scattering is intimately connected with the soft-mode energy. Cochran[22] has predicted the introduced intensity to be proportional to the factor $T(T-T_c)^{-1}$ in a temperature region above the transition point. This scattering has actually been measured on a single potassium dideuterium phosphate, KD_2PO_4, by Buyers *et al.*,[23] leading us to the conclusion that the present theory seems to agree with the experimental observation. Nevertheless, in order to test the present theory completely it would be extremely valuable if a similar measurement had also been performed on a single potassium dihydrogen phosphate, KH_2PO_4.

It is a common fact that the dielectric susceptibility of a given optical medium is closely connected with the optical branches in the following way, see references 24–28,

$$\frac{\varepsilon(\infty)}{\varepsilon(0)} = \frac{\omega_T^2}{\omega_L^2}, \tag{4.85}$$

where T and L refer to the limiting transverse and longitudinal frequencies, respectively. Here $\varepsilon(0)$ and $\varepsilon(\infty)$ designate the static dielectric constant and the square of the optical refractive index, respectively. The former quantity is of a particular importance within the present subject; for if an anomalously huge value for $\varepsilon(0)$ should characterize a given dielectric substance then equation (4.85) would hold only provided that the transverse optical frequency vanishes, ω_T tends to zero. Indeed, the present pseudo-spin method provides a key equation which meets the preceding requirement. Writing equation (4.84) in the form

$$E^2(q) = P(T-T_c)+Qq^2+\ \ldots \tag{4.86}$$

with

$$P = \lambda(2\zeta SJ)^2,$$
$$Q = \zeta(2aSJ)^2$$

we see immediately that the energy of a characteristic normal vibration (also named the "tunneling" mode) is essentially identical with the energy corresponding to the transverse optical frequency, $\hbar\omega_T$. It is interesting to note that the above equation is also obtained by Blinc[29] and by Pytte and Thomas[30] using a slightly different mathematical vocabulary. However, fundamental physical implications coming from the cited papers are identical within the present model.

The Ising model with a transverse field in a high-temperature approximation and at the ground state is studied by Elliott *et al.*[31, 32] in two exhaustive papers where the kinetic energy of the particles, 2Ω, is treated as an independent variable with respect to the effective interaction, $J(0)$. Unlike the present work their critical value of the field, $\Omega_c/J(0)$, depends on the lattice structure and on the interaction range.

A soft mode as associated with the pseudo-spin model is considered by Villain and Aubry[33] by imposing an exact Slater's condition on the hydrogen bondings. However, they find that the soft mode on a neutron scattering pattern corresponds to a pseudo-spin diffusion inelastic and incoherent mode rather than to an optical well-defined phonon.

We conclude the present section by making the following remark: no exact solution to a three-dimensional Ising problem with a transverse field exists as yet, so any result depends heavily on the selected approach. It is possible, however, that an extension of the exact solution on a two-dimensional Ising model with a transverse field which was achieved by several people independently and simultaneously (see references 34–38) may lead us to the eventual goal.

4.9. High-temperature dielectric susceptibilities

Ferroelectric properties of a ferroelectric disappear above the transition temperature, T_c. The transition from the ferroelectric phase to the nonferroelectric one is associated with anomalies in physical properties. For a first-order transition there will be a latent heat, for a second-order one there will be a discontinuity in the specific heat. The spontaneous polarization in the ferroelectric state is associated with spontaneous electrostrictive strains in the crystal. Therefore the ferroelectric structure has a lower symmetry than the paraelectric one.

The dielectric constant ε of a ferroelectric above the transition temperature obeys the Curie–Weiss law

$$\varepsilon = \frac{C'}{T-T_0}+\varepsilon(0) \tag{4.87}$$

where C' is a constant and T_0 is a characteristic temperature which is usually several degrees smaller than the transition temperature, T_c.

Here $\varepsilon(0)$ is a constant whose magnitude depends on the contribution from the electronic polarization. In the vicinity of the transition temperature $\varepsilon(0)$ may be neglected since $\varepsilon \gg \varepsilon(0)$ and $\varepsilon(0)$ is of the order of 1. The dielectric susceptibility is defined by

$$\chi = \frac{\varepsilon - 1}{4\pi} \cong \frac{\varepsilon}{4\pi}. \tag{4.88}$$

In the vicinity of the critical point we may write

$$\chi = \frac{C}{T - T_0} \tag{4.89}$$

where $C = C'/(4\pi)$ is called the Curie constant. For crystals having a KH_2PO_4 structure this constant is in the range of several hundred absolute degrees.

From a theoretical point of view the following problem is very important. Can the present model explain the Curie–Weiss law using the general methods of thermodynamics? To solve this problem we proceed as follows. Let us introduce a canonical ensemble, as done in reference 39, which is the local equilibrium ensemble in the present theory to be derived from the hamiltonian

$$\mathcal{H}(t) = \mathcal{H} + G \sum_k h(\boldsymbol{k})\,(n_k - \langle n_k \rangle), \tag{4.90}$$

where $\mathcal{H}$ and n_k are determined as usual. The quantity $h(\boldsymbol{k})$ represents a thermodynamic electric field which arises as soon as the system of ferroelectric pseudo-spins deviates from equilibrium. The nonequilibrium state is characterized by a spontaneous polarization $P(\boldsymbol{k}, t)$ where $\boldsymbol{k}$ designates a reciprocal lattice vector to be identified with the linear momentum in units $\hbar = 1$ of an external electromagnetic field, whereas t is the local time. Here G designates the fundamental coupling constant expected to have a magnitude of the dipole moment of the complex $K^+(H_2PO_4)^-$. (In the field of magnetism G may be identified with the quantity $g\mu_B$.) The thermal average $\langle n_k \rangle$ is written

$$\langle n_k \rangle = \frac{1}{\exp\,[\beta E(\boldsymbol{k})] - 1} \tag{4.91}$$

where $E(\boldsymbol{k})$ is the energy of the ferroelectric mode, whereas $\beta = (k_B T)^{-1}$. The above expression gives an equilibrium number of quanta, so any departure from the equilibrium state may be treated as a weak perturbation. Therefore the second term in (4.90) can be considered as a small correction to the principal hamiltonian which represents the state of equilibrium.

Next we assume that the thermodynamic electric field can be obtained to a first-order approximation from the principal hamiltonian by differentiating it with respect to the equilibrium number of quanta,

$$Gh(\boldsymbol{k}) = \left.\frac{\partial \mathcal{H}}{\partial n_{\boldsymbol{k}}}\right|_{n_{\boldsymbol{k}} = \langle n_{\boldsymbol{k}} \rangle} = E(\boldsymbol{k}) + \left.\frac{\partial \mathcal{H}_A}{\partial n_{\boldsymbol{k}}}\right|_{n_{\boldsymbol{k}} = \langle n_{\boldsymbol{k}} \rangle} \tag{4.92}$$

So the perturbation hamiltonian appears as an expansion term around the equilibrium value.

In general the thermodynamic electric field is a vector given by the components

$$\boldsymbol{h}(k,\, t) = \{h_a, h_b, h_c\}, \tag{4.93}$$

where a, b, c denote the crystallographic axes. In particular, the c component $h_c = h(\boldsymbol{k}, t)$ is a function of the reciprocal lattice as well as of the local field and can be identified with $h(\boldsymbol{k})$ at $t = 0$ appearing in the expansion (4.90). Also the spontaneous polarization is a vector having the components

$$\boldsymbol{P}(\boldsymbol{k},\, t) = \{P_a, P_b, P_c\}. \tag{4.94}$$

Only the last component can be calculated using the present model.

The generalized susceptibility appears as a tensor with nine independent components at most,

$$\chi(\boldsymbol{k}) = \{\chi_{a'b'}\}, \qquad a',\, b' = a,\, b,\, c. \tag{4.95}$$

The susceptibility tensor can be written

$$\boldsymbol{P}(\boldsymbol{k},\, t) = \hat{\chi}(\boldsymbol{k}) \otimes \boldsymbol{h}(\boldsymbol{k},\, t), \tag{4.96}$$

where $\otimes$ stands for a tensor multiplication. In crystals having a tetragonal symmetry like KH_2PO_4 and isomorphous compounds a rotational symmetry appears with respect to the c-axis. Therefore

$$h_a = h_b, \qquad P_a = P_b,$$

$$\hat{\chi} = \begin{bmatrix} \chi_{aa} & \chi_{ab} & \chi_{ac} \\ \chi_{ab} & \chi_{aa} & \chi_{ac} \\ \chi_{ac} & \chi_{ac} & \chi_{cc} \end{bmatrix}. \tag{4.97}$$

Hence the susceptibility tensor has only four independent components at most, those being χ_{aa}, χ_{ab}, χ_{ac}, and χ_{cc}. If we further assume that no spontaneous polarization is observed in the ab plane then there follows

$$h_a = P_a = 0, \qquad \chi_{ac} = 0, \qquad P_c(\boldsymbol{k}, t) = \chi h_c(\boldsymbol{k}, t), \tag{4.98}$$

where $\chi \equiv \chi_{cc}$. The component χ_{cc} can be calculated explicitly within the present model while the remaining two components, χ_{aa} and χ_{ab}, cannot be done without an essential improvement of the model.

Introduce the density operator

$$\varrho(t) = Z^{-1} \exp[-\beta\mathscr{H}(t)], \tag{4.99}$$

where Z denotes the normalization factor to be determined from the condition

$$\operatorname{Tr} \varrho(t) = 1. \tag{4.100}$$

If we define the density operator for the equilibrium state

$$\varrho = \frac{\exp(-\beta\mathscr{H})}{\operatorname{Tr}\exp(-\beta\mathscr{H})}$$

and take that $h(\boldsymbol{k})$ is much smaller than $\mathscr{H}$ then $\varrho(t)$ can be expanded around the value $n_{\boldsymbol{k}} = \langle n_{\boldsymbol{k}} \rangle$. Indeed, let us start from the operator

$$\hat{G}(\beta) = e^{\beta\mathscr{H}} e^{-\beta\mathscr{H}(t)} \tag{4.101}$$

where $\mathscr{H}(t)$ is given by (4.90). Clearly $G(0) = 1$. Also we have

$$\frac{d\hat{G}(\beta)}{d\beta} = -e^{\beta\mathscr{H}} \hat{h} e^{-\beta\mathscr{H}} \hat{G}(\beta), \tag{4.102}$$

where

$$\hat{h} = G \sum_{k} h(\mathbf{k})\,(n_k - \langle n_k \rangle). \tag{4.103}$$

Using the iteration method we obtain

$$\hat{G}(\beta) = \hat{G}(0) - \int_0^\beta \exp(\lambda\mathcal{H})\,\hat{h}\,\exp(-\lambda\mathcal{H})\,\hat{G}(\beta)\,d\lambda$$

$$= \hat{G}(0) - \int_0^\beta \exp(\lambda\mathcal{H})\,\hat{h}\exp(-\lambda\mathcal{H})\,\hat{G}(0)\,d\lambda + \dots \tag{4.104}$$

where higher-order terms are neglected. By definition (4.99) we can write

$$\varrho(t) = Z^{-1}\exp(-\beta\mathcal{H})\,\hat{G}(\beta)$$

$$= \varrho\left[1 - \int_0^\beta \exp(\lambda\mathcal{H})\,\hat{h}\,\exp(-\beta\mathcal{H})\,d\lambda\right]. \tag{4.105}$$

Next we define the c-component of the spontaneous polarization in the form

$$P_c(\mathbf{k}, t) = NG\,\mathrm{Tr}\,\varrho(t)\,n_k \tag{4.106}$$

where $\varrho(t)$ is given by (4.105), G is the fundamental coupling constant, and N is the total number of dipoles per unit volume. Using the previous equations we arrive at the result

$$\chi^{-1}(\mathbf{k}) = G^{-2}\left.\frac{\partial\mathcal{H}}{\partial n_k}\right|_{n_k=\langle n_k\rangle}[N\,\mathrm{Tr}\,\varrho(t)\,n_k]^{-1}$$

where

$$\mathrm{Tr}\,\varrho(t)\,n_k = \langle n_k\rangle + G\sum_{k'} h(\mathbf{k}')\,\mathrm{Tr}\left[\varrho\int_0^\beta \exp(\lambda\mathcal{H})\,n_{k'}\right.$$

$$\left.\times\exp(-\lambda\mathcal{H})\,d\lambda\right]. \tag{4.107}$$

Neglecting higher-order terms in (4.107) leads to

$$\chi^{-1}(\mathbf{k}) = N^{-1}G^{-2}k_B T_c[P(T-T_c)+Qk^2]. \tag{4.108}$$

As $\boldsymbol{k}$ tends to zero one can write the susceptibility in the form

$$\chi(0) = \frac{C}{T-T_c}, \qquad T \geqslant T_c,$$
$$C = \frac{NG^2k_B}{P} T_c \tag{4.109}$$

where C may be identified with the Curie constant. Therefore the Curie–Weiss law, equation (4.89), can be formulated from very general arguments within the present model.

For a hydrogen-bonded ferroelectric we postulate the fundamental coupling constant, G, following Uehling,[40]

$$G = 2\mu[K^+(H_2PO_4)^-] \tag{4.110}$$

where μ designates the dipole moment of the indicated ions. Writing the saturation polarization as

$$P_s = N\mu = \frac{2G}{a^2c} \tag{4.111}$$

one can estimate the coupling constant by comparing the above expression to the observed polarization.

The energy of the ferroelectric mode and the Curie–Weiss law on the paraelectric side can be written

$$E^2(\boldsymbol{q}) = P(T-T_c)+Qq^2$$
$$\chi = \frac{C_p}{T-T_c} \tag{4.112}$$

with the expansion coefficients

$$P = \lambda_p H_0^2,$$
$$C_p = \frac{NG^2k_B}{P} T_c, \tag{4.113}$$

H_0 designates the zero-temperature value for the molecular field (see Chapter 2).

On the other hand, the expansion coefficient on the ferroelectric side, λ_f, can be evaluated as follows. Let us write the molecular-field equation in the form

$$H = H_c[1+\lambda_f(T_c-T)] \tag{4.114a}$$

where

$$H_c = \sqrt{(1-\Theta_c^2)}\, H_0$$
$$\Theta_c = 2k_B T_c/H_0. \tag{4.114b}$$

There follows the relation

$$\lambda_f \cong \frac{\Theta_c^2}{T_c(1-\Theta_c^2)}\,. \tag{4.115}$$

Important numerical analysis using an idealized lattice structure is presented in Table 4.7 and Table 4.8 on the characteristic parameters for both sides of the critical temperature. It is worth noting that the presented analysis is not particularly sensitive on the selected idealized

TABLE 4.7.

Coupling Constant, G, Paraelectric Curie Constant, C_p, and Expansion Coefficients, P and λ_p, for Hydrogen-bonded Ferroelectrics

T_c, K	G elstcgs units	C_p observed	P $(\text{cm}^{-1})^2$/ degree	H_0 ‖ cm^{-1}	λ_p $(\text{degree})^{-1}$
			KH_2PO_4		
122	1.37×10^{-18}	258.6‡	32.22	383 357	2.20×10^{-4} 2.53×10^{-4}
			KD_2PO_4		
223	1.94×10^{-18}	334.2§	91.14	380 360	6.31×10^{-4} 7.03×10^{-4}

‡ After Jona and Shirane, cited in Chapter 2. § After Hill and Ichiki.[41]
‖ Present method using data in Chapter 2.

TABLE 4.8.

Dimensionless Transition Temperature, Θ_c, Expansion Coefficient, λ_f, and Ratios λ_f/λ_p and C_p/C_f for Hydrogen-bonded Ferroelectrics

Θ_c	λ_f (degree)$^{-1}$	λ_f/λ_p calculated	C_p/C_f calculated	C_p/C_f observed
		KH_2PO_4		
0.443	2.00×10^{-3}	9.1	9.1	~ 10‡
0.475	2.39×10^{-3}	9.5	9.5	
		KD_2PO_4		
0.816	8.92×10^{-3}	14.1	14.1	~ 16§
0.861	12.86×10^{-3}	18.3	18.3	

‡ After Jona and Shirane, cited in Chapter 2. § After Hill and Ichiki.[41]

lattice structure. The inverse dielectric susceptibilities are illustrated in Fig. 4.2. Here one can observe a huge isotope effect on the slope of the inverse susceptibility as viewed against temperature.

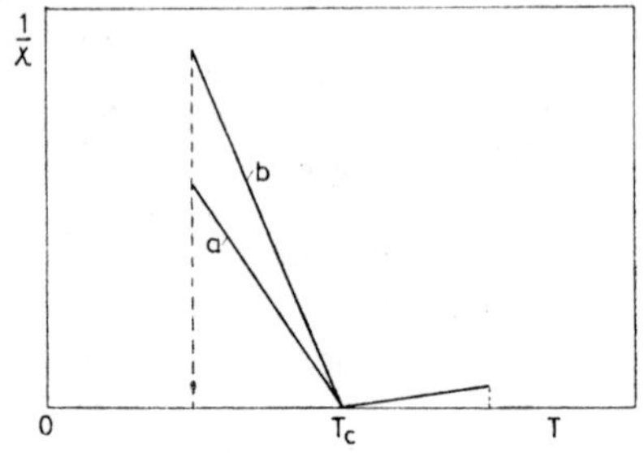

FIG. 4.2. Schematic representation of the inverse dielectric susceptibility in the vicinity of the transition temperature: (a) for a hydrogen-bonded ferroelectric; (b) for a deuterium-bonded ferroelectric. A huge isotope effect on the slope of the curve is easily observed.

References

1. R. KUBO, *J. Phys. Soc. Japan* **12,** 570 (1957); R. KUBO, M. YOKOTA, and S. NAKAJIMA, *ibid.* **12,** 1203 (1957).
2. D. N. ZUBAREV, *Usp. Fiz. Nauk* **3,** 320 (1960) [English trans.: *Soviet Phys. — Usp.* **71,** 71 (1960)].
3. S. V. TYABLIKOV, *Methods in the Quantum Theory of Magnetism*, Plenum Press, Inc., New York, 1967; S. V. TYABLIKOV, *Dokl. Akad. Nauk USSR* **149,** 573 (1963).
4. J. H. BARRY, *Phys. Rev.* **174,** 531 (1968).
5. R. P. KENAN, *J. Appl. Phys.* **37,** 1453 (1966); R. P. KENAN, *Phys. Rev.* **159,** 430 (1967).
6. P. GOSAR, *Fizika (Zagreb)* **1,** 147 (1969).
7. H. STERN, *Phys. Rev.* **147,** 94 (1966).
8. G. WICK, *Phys. Rev.* **80,** 268 (1950).
9. C. BLOCH and C. DE DOMINICIS, *Nucl. Phys.* **7,** 459 (1958).
10. H. B. CALLEN, in: *Physics of Many-particle Systems*, Vol. I, ed. by E. MEERON, Gordon & Breach, New York, 1966.
11. W. MARSHALL, *Proc. Roy. Soc.* (London) A, **232,** 48, 69 (1955).
12. A. A. ABRIKOSOV, L. P. GORKOV, and I. Y. DZYALOSHINSKII, *Quantum Field Theoretical Methods in Statistical Physics*, Pergamon Press, Oxford, 1965.
13. P. W. ANDERSON, *Phys. Rev.* **86,** 694 (1952).
14. R. KUBO, *Phys. Rev.* **87,** 568 (1952).
15. S. T. DEMBINSKI, *Can. J. Phys.* **46,** 1435 (1968).
16. D. J. THOULESS, *Proc. Phys. Soc.* **90,** 243 (1967).
17. M. NAUCIEL-BLOCH, *Ann. Phys.* (Paris) **5,** 139 (1970).
18. O. NAGAI and T. TANAKA, *Phys. Rev.* **188,** 821 (1969).
19. M. RUBINSTEIN and V. J. FOLEN, *Phys. Lett.* **28** A, 108 (1968).
20. B. O. LOOPSTRA, B. VAN LAAR, and D. J. BREED, *Phys. Lett.* **26** A, 526 (1968).
21. R. A. COWLEY, *Phys. Rev.* **134,** A 981 (1964).
22. W. COCHRAN, *Adv. Phys. (Phil. Mag. Suppl.)* **18,** 157 (1969).
23. W. J. L. BUYERS, R. A. COWLEY, G. L. PAUL, and W. COCHRAN, in: *Neutron Inelastic Scattering*, Vol. I, International Atomic Energy Agency, Vienna, 1968.
24. R. H. LYDDANE, R. G. SACHS, and E. TELLER, *Phys. Rev.* **59,** 673 (1941).
25. M. BORN and K. HUANG, *Dynamical Theory of Crystal Lattices*, Oxford University Press, 1954.
26. W. COCHRAN, *Z. Kristallogr.* **112,** 465 (1959).
27. W. COCHRAN and R. A. COWLEY, *J. Phys. Chem. Solids* **23,** 447 (1962).
28. A. S. BARKER, *Phys. Rev.* **136,** A 1290 (1964).
29. R. BLINC, in: *Theory of Condensed Matter*, International Atomic Energy Agency, Vienna, 1968.
30. E. PYTTE and H. THOMAS, *Phys. Rev.* **175,** 610 (1968).

31. R. J. Elliot and C. Wood, *J. Phys. C (Solid State Phys.)* **4,** 2359 (1971).
32. P. Pfeuty and R. J. Elliott, *J. Phys. C (Solid State Phys.)* **4,** 2370 (1971).
33. J. Villain and S. Aubry, *Phys. Stat. Solidi* **33,** 337 (1969).
34. F. Y. Wu, *Phys. Rev. Lett.* **18,** 605 (1967).
35. E. H. Lieb, *Phys. Rev. Lett.* **19,** 108 (1967); E. H. Lieb, in: *Contemporary Physics*, Vol. I, Trieste Symposium 1968, International Atomic Energy Agency, Vienna, 1969.
36. B. Sutherland, *Phys. Rev. Lett.* **19,** 103 (1967).
37. C. P. Yang, *Phys. Rev. Lett.* **19,** 586 (1967).
38. B. Sutherland, C. N. Yang, and C. P. Yang, *Phys. Rev. Lett.* **19,** 588 (1967).
39. L. Novaković, S. Stamenković, and A. Vlahov, *J. Phys. Chem. Solids* **32,** 487 (1971).
40. E. A. Uehling, Theories of ferroelectricity in KH_2PO_4, in: *Lectures in Theoretical Physics*, Vol. V, ed. by W. E. Brittin, B. W. Downs, and J. Downs, Interscience Publishers, John Wiley & Sons, Inc., New York, 1963.
41. R. M. Hill and S. K. Ichiki, *Phys. Rev.* **132,** 1603 (1963).

APPENDIX A

Cubic Lattices

Basic features of the simple (P), body-centered (I), and face-centered (F) cubic lattices are given in Table A1. Using these data the following sum is readily calculated

$$K(\boldsymbol{q}) = \frac{1}{z}\sum_{k=1}^{z} K_{jk} \exp\left([i\boldsymbol{q}\cdot(\boldsymbol{R}_j - \boldsymbol{R}_k)\right]$$

$$= \frac{1}{z}\sum_{k=1}^{z} K_{jk} \cos\left[\boldsymbol{q}\cdot(\boldsymbol{R}_j - \boldsymbol{R}_k)\right] \qquad \text{(A1)}$$

Table A1.

Basic Features of the Cubic Lattices

Lattice	P	I	F
Unit-cell volume	a^3	a^3	a^3
Nearest-neighbor distance	a	$a\sqrt{(3)}/2$	$a/\sqrt{2}$
The number of nearest neighbors, z	6	8	12
The number of sites per unit cells	1	2	4

where

$$\boldsymbol{R}_j = 0, \qquad \boldsymbol{R}_k = \{X_k, Y_k, Z_k\}. \qquad \text{(A2)}$$

The reciprocal lattice vector is defined by

$$\begin{aligned} \boldsymbol{q} &= \{q_x, q_y, q_z\}, \\ q_x &= q \cos\varphi \sin\vartheta, \\ q_y &= q \sin\varphi \sin\vartheta, \\ q_z &= q \cos\vartheta. \end{aligned} \tag{A3}$$

Nearest-neighbor distances are given in Tables A2–4.

TABLE A2.

Nearest-neighbor Distances on a Cubic P *Lattice in Units a* (Fig. A1)

k	1	2	3	4	5	6
X	1	−1	0	0	0	0
Y	0	0	1	−1	0	0
Z	0	0	0	0	1	−1

TABLE A3.

Nearest-neighbor Distances on a Cubic I *Lattice in Units a*/2 (Fig. A2)

k	1	2	3	4	5	6	7	8
X	1	1	−1	−1	1	1	−1	−1
Y	−1	1	1	−1	−1	1	1	−1
Z	1	1	1	1	−1	−1	−1	−1

TABLE A4.

Nearest-neighbor Distances on a Cubic F *Lattice in Units a*/2 (Fig. A3)

k	1	2	3	4	5	6	7	8	9	10	11	12
X	1	1	−1	−1	0	0	0	0	1	1	−1	−1
Y	1	−1	1	−1	1	1	−1	−1	0	0	0	0
Z	0	0	0	0	1	−1	1	−1	1	−1	1	−1

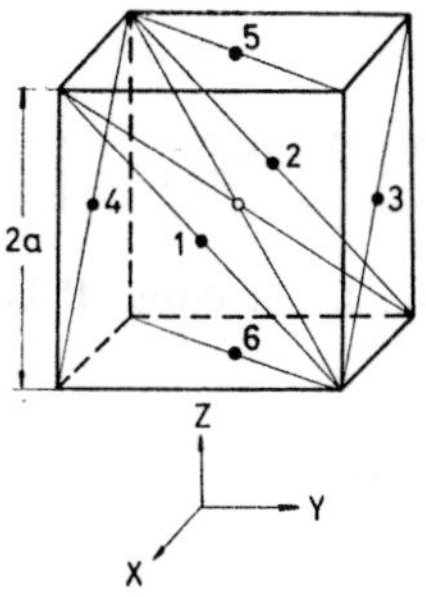

Fig. A1.

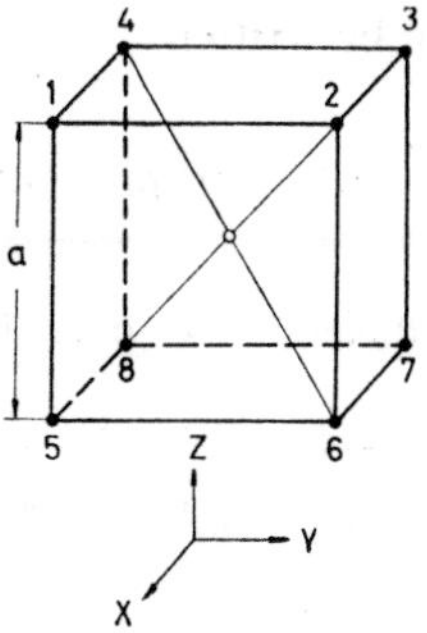

Fig. A2.

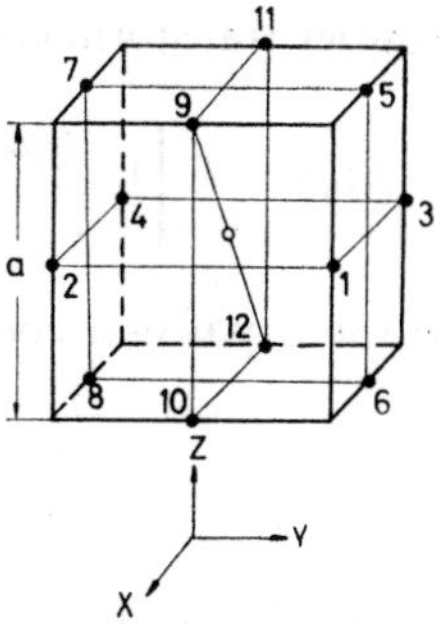

Fig. A3.

Fig. A1. A given atom with six nearest neighbors on the simple cubic lattice.

Fig. A2. A given atom with eight nearest neighbors on the body-centered cubic lattice.

Fig. A3. A given atom with twelve nearest neighbors on the face-centered cubic lattice.

The Fourier transform is expressed as follows: for a cubic P lattice,

$$K(\boldsymbol{q}) = \frac{K}{3}\,[\cos(q_x a)+\cos(q_y a)+\cos(q_z a)]; \tag{A4}$$

for a cubic I lattice,

$$K(\boldsymbol{q}) = K\cos\frac{q_x a}{2}\cos\frac{q_y a}{2}\cos\frac{q_z a}{2}; \tag{A5}$$

for a cubic F lattice,

$$K(\boldsymbol{q}) = \frac{K}{3}\left[\cos\frac{q_x a}{\sqrt{2}}+\cos\frac{q_y a}{\sqrt{2}}+\cos\frac{q_z a}{\sqrt{2}}\right]. \tag{A6}$$

Introducing the abbreviations

$$\cos\varphi = \alpha,$$

$$\sin\varphi = \beta,$$

$$\cos\vartheta = \gamma,$$

$$\sin\vartheta = \delta,$$

the Fourier transform can be written

$$K(\boldsymbol{q}) = K\left\{1-\frac{(qa)^2}{z}+A(\varphi,\vartheta)\,(qa)^4-B(\varphi,\vartheta)\;(qa)^6+\ldots\right\} \quad \text{(A7)}$$

where the expansion coefficients are given by, for a cubic P lattice,

$$\begin{aligned} A &= \tfrac{1}{72}\{(\alpha\delta)^4+(\beta\delta)^4+\gamma^4\}, \\ B &= \tfrac{1}{2160}\{(\alpha\delta)^6+(\beta\delta)^6+\gamma^6\}; \end{aligned} \quad \text{(A8)}$$

for a cubic I lattice,

$$\begin{aligned} A &= \tfrac{1}{384}\{(\alpha\delta)^4+(\beta\delta)^4+\gamma^4\}+\tfrac{1}{64}\{(\gamma\delta)^2+(\alpha\beta)^2\,\delta^4\}, \\ B &= \frac{1}{2^6\cdot 6!}\{(\alpha\delta)^6+(\beta\delta)^6+\gamma^6\} \\ &\quad + \frac{1}{2^7\cdot 4!}\{\alpha^2\beta^4\,\delta^6+(\alpha\delta)^2\gamma^4+(\beta\delta)^2\,\gamma^4 \\ &\quad + \alpha^4\beta^2\delta^6+\alpha^4\gamma^2\delta^4+\beta^4\gamma^2\delta^4\} \\ &\quad + \frac{1}{2^8\cdot 2!}(\alpha\beta\gamma)^2\,\delta^4; \end{aligned} \quad \text{(A9)}$$

for a cubic F lattice,

$$\begin{aligned} A &= \tfrac{1}{288}\{(\alpha\delta)^4+(\beta\delta)^4+\gamma^4\}, \\ B &= \tfrac{1}{17\,280}\{(\alpha\delta)^6+(\beta\delta)^6+\gamma^6\}. \end{aligned} \quad \text{(A1)}$$

APPENDIX B

A Method Used to Separate the Proton Coordinates from Those of Heavy Clusters

THE total hamiltonian consists of kinetic and potential energies referring to the protons and heavy clusters in a given one-dimensional hydrogen-bonded ferroelectric as follows:

$$
\begin{aligned}
\mathcal{H} &= T+U,\\
U &= U_0+U_1,\\
U_0 &= K\sum_n u_{2n+1}^2 + A\sum_n u_{2n+1}(S_{2n}^z+S_{2n+2}^z),\\
T &= \frac{1}{2M}\sum_n p_{2n+1}^2 - 2\Omega\sum_n S_{2n}^x - B\sum_n S_{2n}^z S_{2n+2}^x,\\
U_1 &= -C\sum_n S_{2n}^z S_{2n+2}^z. \qquad \text{(B1)}
\end{aligned}
$$

Clearly the total hamiltonian is invariant under the simultaneous operations

$$
\begin{aligned}
u_{2n+1} &\to -u_{2n+1},\\
S_{2n}^z &\to -S_{2n}^z
\end{aligned}
$$

which follow from symmetry arguments. By the substitution

$$u_{2n+1} = v_{2n+1}+\xi_{2n+1} \qquad \text{(B2)}$$

the term U_0 becomes transformed

$$U_0 = K\sum_n v_{2n+1}^2 + \sum_n v_{2n+1}\{2K\xi_{2n+1} + A(S_{2n}^z + S_{2n+1}^z)\}$$
$$+ \sum_n \{K\xi_{2n+1}^2 + A(S_{2n}^z + S_{2n+1}^z)\,\xi_{2n+1}\}. \tag{B3}$$

We demand for the second sum in (B3) to vanish identically which leads to

$$\xi_{2n+1} = -\frac{A}{2K}(S_{2n}^z + S_{2n+1}^z).$$

Hence,

$$U_0 = -\frac{A^2}{2K} + K\sum_n v_{2n+1}^2 - \frac{A^2}{2K}\sum_n S_{2n}^z S_{2n+2}^z.$$

Now the total hamiltonian becomes

$$\mathcal{H} = \mathcal{H}_0 + \mathcal{H}_1$$

where

$$\mathcal{H}_0 = \frac{1}{2M}\sum_n p_{2n+1}^2 + K\sum_n v_{2n+1}^2 \tag{B4}$$

$$-\mathcal{H}_1 = 2\Omega\sum_n S_{2n}^x + \sum_n (BS_{2n}^x S_{2n+2}^x + JS_{2n}^z S_{2n+2}^z). \tag{B5}$$

Here, J designates an effective proton–proton exchange integral,

$$J = C + \frac{A^2}{2K} \tag{B6}$$

Therefore the actual proton–lattice interaction term as given by the second sum in U_0 is written by the pseudo-spin components as part of the effective proton–proton exchange integral.

APPENDIX C

The Use of Gamma and Zeta Functions

IN THE present analysis we use integrals of the form

$$I_n(\tau) = \int_0^\infty \frac{q^{2n+2}\,dq}{\exp\left[(1-\eta)\dfrac{(aq)^2}{\tau}\right]-1} \tag{C1}$$

where $n = 0, 1, 2, \ldots$, η is some parameter such that $\eta \geqslant 0$, τ is a dimensionless temperature defined by

$$\tau = \frac{kT}{2SJ}\,. \tag{C2}$$

By the substitution

$$(1-\eta)\frac{(aq)^2}{\tau} = x$$

we obtain

$$I_n(\tau) = \tfrac{1}{2}[a^2(1-\eta)]^{-(n+3/2)}\,\tau^{n+3/2}\,K_n, \tag{C3}$$

$$K_n = \int_0^\infty \frac{x^{n+1/2}\,dx}{\exp x - 1}\,. \tag{C4}$$

The latter integral is readily calculated

$$K_n = \Gamma(n+\tfrac{3}{2})\,\zeta(n+\tfrac{3}{2}) \tag{C5}$$

where Γ and ζ designate the well-known gamma function and Riemann zeta function, respectively. By definition

$$\Gamma\left(n+\frac{3}{2}\right) = 2\int_0^\infty x^{2(n+3/2)-1}\exp(-x^2)\,dx = \frac{(2n+1)!!}{2^{n+1}}\sqrt{\pi}, \tag{C6}$$

if n is integer. For any number

$$\Gamma(\nu+1) = \nu\Gamma(\nu);$$

and if ν is integer

$$\Gamma(\nu+1) = \nu!.$$

The Riemann zeta function is defined by

$$\zeta(\nu) = [\Gamma(\nu)]^{-1}\int_0^\infty \frac{x^{\nu-1}\exp(-x)\,dx}{1-\exp(-x)} = \sum_{m=1}^{\infty}\frac{1}{m^\nu}, \qquad \nu > 1. \tag{C7}$$

The first several values of these functions for various arguments often used in the present analysis are given in Tables C1 and C2.

TABLE C1.

The Gamma Function

ν	$\frac{1}{2}$	1	$\frac{3}{2}$	2	$\frac{5}{2}$	3	$\frac{7}{2}$
$\Gamma(\nu)$	$\sqrt{\pi}$	1	$\sqrt{(\pi)}/2$	1	$3\sqrt{(\pi)}/4$	2	$15\sqrt{(\pi)}/8$

TABLE C2.

The Riemann Zeta Function

ν	$\frac{3}{2}$	2	$\frac{5}{2}$	3	$\frac{7}{2}$	4	$\frac{9}{2}$
$\zeta(\nu)$	2.612	1.645	1.341	1.202	1.127	1.082	1.055

Also we use integrals of the form

$$J_n(\tau) = \int_0^\infty dq \int_0^{2\pi} d\varphi \int_0^\pi \sin\vartheta \, d\vartheta \frac{q^{2n+2}}{\exp x - 1}, \tag{C8}$$

where

$$x = \frac{1}{\tau}[(1-\eta)(qa)^2 - zA(\varphi, \vartheta)(qa)^4 + zB(\varphi, \vartheta)(qa)^6 + \dots]. \tag{C9}$$

Here, $n = 0, 1, 2, \dots$; $A(\varphi, \vartheta)$ and $B(\varphi, \vartheta)$ are evaluated in Appendix A.

Introduce the substitution

$$x = C(1 - Dq^2 + Eq^4)\, q^2, \tag{C10}$$

$$C = \frac{1}{\tau}(1-\eta)\, a^2,$$

$$D = \frac{zA(\varphi, \vartheta)\, a^2}{1-\eta},$$

$$E = \frac{zB(\varphi, \vartheta)\, a^4}{1-\eta}$$

to obtain from equation (C10)

$$Cq^2 = \frac{x}{1-(Dq^2 - Eq^4)} \cong x[1 + Dq^2 - Eq^4 + (Dq^2 - Eq^4)^2]. \tag{C11}$$

The above expansion holds only for small values of x as indicated by the symbol of an approximate equality. The first few terms are

$$Cq^2 \cong x\left(1 + \frac{D}{C}x + \frac{2D^2 - E}{C^2}x^2 + \dots\right)$$

$$= x\{1 + zA(\varphi, \vartheta)(1-\eta)^{-2}\tau x + [2z^2A^2(\varphi, \vartheta)(1-\eta)^{-4}$$

$$- zB(\varphi, \vartheta)(1-\eta)^{-3}](\tau x)^2 + \dots\}. \tag{C12}$$

There follows the result of integration for the integral J_0, equation (C8),

$$J_0(\tau) = \frac{1}{2} C^{-3/2} \int_0^{2\pi} d\varphi \int_0^{\pi} \sin\vartheta \, d\vartheta \int_0^{\infty} \frac{\sqrt{x}\, f(x)\, dx}{\exp x - 1}$$

where

$$f(x) = 1 + \tfrac{5}{2} z A(\varphi, \vartheta)(1-\eta)^{-2}\tau x + [8z^2 A^2(\varphi, \vartheta)(1-\eta)^{-4} - \tfrac{7}{2} z B(\varphi, \vartheta)(1-\eta)^{-3}](\tau x)^2 + \dots \quad \text{(C13)}$$

The integration over the angles φ and ϑ can be done immediately by making use of the following expressions:

$$\int_0^{\pi/2} (\sin x)^{m-1}\, dx = \int_0^{\pi/2} (\cos x)^{m-1}\, dx = 2^{m-2} B\left(\frac{m}{2}, \frac{m}{2}\right),$$

$$\int_0^{\pi/2} (\sin x)^{2m}\, dx = \int_0^{\pi/2} (\cos x)^{2m}\, dx = \frac{(2m-1)!!}{2(2m)!!}\pi,$$

$$\int_0^{\pi/2} (\sin x)^{2m+1}\, dx = \int_0^{\pi/2} (\cos x)^{2m+1}\, dx = \frac{(2m)!!}{(2m+1)!!},$$

$$\int_0^{\pi/2} (\sin x)^{m-1} (\cos x)^{n-1}\, dx = \frac{1}{2} B\left(\frac{m}{2}, \frac{n}{2}\right)$$

where the so-called beta function is defined by

$$B\left(\frac{m}{2}, \frac{n}{2}\right) = \frac{\Gamma\left(\frac{m}{2}\right)\Gamma\left(\frac{n}{2}\right)}{\Gamma\left(\frac{m+n}{2}\right)}.$$

TABLE C3.

The Result of Integration Over the Angles φ and ϑ for Cubic Lattices

	L_1	L_2	L_3
P	$\frac{\pi}{30}$	$41\frac{\pi}{136\,080}$	$\frac{\pi}{1\,260}$
I	$3\frac{\pi}{160}$	$353\frac{\pi}{945\times 4\,096}$	$3\frac{\pi}{35\times 256}$
F	$\frac{\pi}{120}$	$41\frac{\pi}{16\times 136\,080}$	$\frac{\pi}{10\,080}$

In Table C3:

$$L_1 = \int_0^{2\pi} d\varphi \int_0^{\pi} A(\varphi, \vartheta) \sin\vartheta \, d\vartheta,$$

$$L_2 = \int_0^{2\pi} d\varphi \int_0^{\pi} A^2(\varphi, \vartheta) \sin\vartheta \, d\vartheta,$$

$$L_3 = \int_0^{2\pi} d\varphi \int_0^{\pi} B(\varphi, \vartheta) \sin\vartheta \, d\vartheta.$$

Therefore by virtue of equations (C4) and (C5) the integral J_0 becomes

$$J_0(\tau) = C^{-3/2}\left\{2\pi\Gamma\left(\frac{3}{2}\right)\zeta\left(\frac{3}{2}\right) + \frac{5z}{4}L_1\Gamma\left(\frac{5}{2}\right)\zeta\left(\frac{5}{2}\right)(1-\eta)^{-2}\tau\right.$$
$$\left.+\left[4z^2L_2(1-\eta)^{-4} - \frac{7z}{4}L_3(1-\eta)^{-3}\right]\Gamma\left(\frac{7}{2}\right)\zeta\left(\frac{7}{2}\right)\tau^2\right\} \quad \text{(C14)}$$

where C is given by (C10). Clearly $J_0(\tau)$ has the following expansion in terms of τ:

$$J_0(\tau) = \frac{\tau^{3/2}}{a^3}\left[b_0 + b_1\tau + b_2\tau^2 + b_3\tau^{5/2} + (\tau^3)\right]$$

where the expansion coefficients b_j depends the type of the lattice considered. As an illustration we quote below the result obtained for cubic lattices having done the indicated integrations,

$$\eta = \frac{\pi^{-3/2}\tau^{5/2}}{32S}\,\zeta\left(\frac{5}{2}\right),$$

$$b_0 = \pi^{3/2}\,\zeta(\tfrac{3}{2}),$$

$$b_1 = \frac{5z}{4}L_1\Gamma\left(\frac{5}{2}\right)\zeta\left(\frac{5}{2}\right),$$

$$b_2 = \left(4z^2L_2 - \frac{7z}{4}L_3\right)\Gamma\left(\frac{7}{2}\right)\zeta\left(\frac{7}{2}\right),$$

$$b_3 = \frac{3}{32S}\pi^{-1/2}\,\Gamma\left(\frac{3}{2}\right)\zeta\left(\frac{3}{2}\right)\zeta\left(\frac{5}{2}\right). \qquad \text{(C15)}$$

TABLE C4.

Expansion Coefficients b_j for Cubic Lattices

	First factor			Second factor
	P	I	F	
b_0	1	1	1	$\pi^{3/2}\,\zeta(\frac{3}{2})$
b_1	$\frac{3}{16}$	$\frac{9}{64}$	$\frac{3}{32}$	$\pi^{3/2}\,\zeta(\frac{5}{2})$
b_2	$\frac{265}{63\times 64}$	$\frac{2\,257}{64\times 1\,008}$	$\frac{265}{63\times 256}$	$\pi^{3/2}\,\zeta\left(\frac{7}{2}\right)$
b_3	$\frac{3}{64S}$	$\frac{3}{64S}$	$\frac{3}{64S}$	$\zeta\left(\frac{3}{2}\right)\zeta\left(\frac{5}{2}\right)$

APPENDIX D

A Method Used to Compute the Sublattice Order Parameter

THE zero-temperature sublattice order parameter, $\sigma(T = 0)$, is derived in Section 4.5. By definition

$$\sigma(T = 0) = 1 - \frac{I-1}{2S}$$

where

$$I = N^{-1} \sum_{p} [1-\gamma^2(\boldsymbol{p})]^{-1/2}. \qquad \text{(D1)}$$

This quantity depends on the type of lattice considered. If a body-centered cubic lattice (CsCl structure) is assumed then there follows‡

$$\gamma(\boldsymbol{p}) = \cos\frac{p_x a}{2} \cdot \cos\frac{p_y a}{2} \cdot \cos\frac{p_z a}{2}$$

$$p_x = \frac{2\pi\nu_1}{aN_1}, \quad p_y = \frac{2\pi\nu_2}{aN_2}, \quad p_z = \frac{2\pi\nu_3}{aN_3},$$

with

$$-\frac{N_i}{2} \leqslant \nu_i \leqslant \frac{N_i}{2} \qquad (i = 1, 2, 3).$$

‡ The cesium-chloride structure belongs to a simple cubic lattice with a body-centered cubic configuration. Each cesium ion is surrounded by eight chlorine ions, and vice versa.

In the limit $N = N_1N_2N_3 \to \infty$ the sum (D1) is readily transformed to the integral

$$I(\text{CsCl}) = \frac{1}{(2\pi)^3} \iiint_{-\pi}^{\pi} [1-(\cos x \cos y \cos z)^2]^{-1/2}\, dx\, dy\, dz.$$

By another transformation

$$I(\text{CsCl}) = \frac{1}{(2\pi)^3} \sum_{\nu=0}^{\infty} (-1)^\nu \beta_\nu \left[\int_{-\pi}^{\pi} (\cos x)^{2\nu}\, dx \right]^3 \tag{D2}$$

with

$$\beta_\nu = \binom{-\frac{1}{2}}{\nu} = (-1)^\nu \frac{(2\nu-1)!!}{(2\nu)!!}$$

$$(-1)!! = 1,$$

$$\int_{-\pi}^{\pi} (\cos x)^{2\nu}\, dx = 2\pi \frac{(2\nu-1)!!}{(2\nu)!!}. \tag{D3}$$

Hence the integral (D2) is reduced to the numerical sum

$$I(\text{CsCl}) = 1 + \sum_{\nu=1}^{N} \left[\frac{(2\nu-1)!!}{(2\nu)!!} \right]^4$$

$$= 1.11863 \tag{D4}$$

where N is equal to 10,000.

If a face-centered cubic lattice (NaCl structure) is assumed then there follows‡

$$\gamma(\boldsymbol{p}) = \tfrac{1}{3}(\cos p_x a + \cos p_y a + \cos p_z a)$$

‡ The sodium-chloride structure belongs to a face-centered cubic lattice with a simple cubic configuration. Each sodium ion is surrounded by six chlorine ions, and vice versa.

with the same definition for the components p_x, p_y, p_z so as to include all the points of the reciprocal lattice space. Now the relevant sum is reduced to the integral

$$I(\text{NaCl}) = \frac{1}{(2\pi)^3} \iiint_{-\pi}^{\pi} \frac{dx\,dy\,dz}{\sqrt{(1-t^2)}} \tag{D5}$$

where

$$t = \tfrac{1}{3}(\cos x + \cos y + \cos z).$$

Using the same transformation as in the previous case we can write

$$\frac{1}{\sqrt{(1-t^2)}} = 1 + \sum_{\lambda=1}^{\infty} (-1)^{\lambda} \beta_{\lambda} 3^{-2\lambda} \sum_{m=0}^{2\lambda} \binom{2\lambda}{m}$$

$$\times \sum_{n=0}^{2\lambda-m} \binom{2\lambda-m}{n} (\cos x)^{2\lambda-m-n} (\cos y)^{n} (\cos z)^{m}. \tag{D6}$$

Having introduced $m = 2\mu$, $n = 2\nu$ we obtain after the indicated integration is performed

$$I(\text{NaCl}) = 1 + \sum_{\lambda=1}^{\infty} (-1)^{\lambda} \beta_{\lambda} 3^{-2\lambda} \sum_{\mu=0}^{\lambda} \binom{2\lambda}{2\mu} |\beta_{\mu}|$$

$$\times \sum_{\nu=0}^{\lambda=\mu} \binom{2\lambda-2\mu}{2\nu} |\beta_{\lambda-\mu-\nu}| \cdot |\beta_{\nu}|$$

$$= 1 + \sum_{\lambda=1}^{N} \left[\frac{(2\lambda)!}{12^{\lambda}.\lambda!}\right]^2 \sum_{\mu=0}^{\lambda} \frac{(2\lambda-2\mu)!}{(\mu!)^2\,[(\lambda-\mu!)]^4}$$

$$= 1.15619, \tag{D7}$$

where N is equal to 250.

Numerical results on the sublattice order parameter $\sigma(T = 0)$ are presented in Table D1 for various values of the sublattice pseudo-spin.

APPENDIX D.

TABLE D1.

Zero-temperature Sublattice Order Parameter for Several Values of the Sublattice Pseudo-spin

Pseudo-spin	NaCl structure	CsCl structure
$\frac{1}{2}$	0.844	0.881
1	0.922	0.941
$\frac{3}{2}$	0.948	0.960
2	0.961	0.970
$\frac{5}{2}$	0.969	0.976
3	0.974	0.980

APPENDIX E

Important Constants and Relations of Atomic Physics

PHYSICAL constants most frequently used in the present and similar works are presented in Table E1, see note‡ at the end of the table.

A great deal of every numerical work is expressed in appropriate energy units. However various energy units are connected by conversion factors which are based on the fundamental relations as follows:

$$\begin{aligned} 1 \text{ electron volt} &= 1.60210\,(2)\times 10^{-19}\ \text{J} \\ &= 1.60210\,(2)\times 10^{-12}\ \text{erg} \\ &= 8065.73\,(8)\ \text{cm}^{-1} \\ &= 2.41804\,(2)\times 10^{14}\ \text{sec}^{-1}. \end{aligned}$$

$$\begin{aligned} &1 \text{ electron volt per particle} \\ &\qquad = 11604.9\,(5)\ \text{K}. \end{aligned}$$

All these conversion factors are collected in Table E2.

TABLE E1.

The Most Frequently used Physical Constants

Constant	Symbol	Numerical value‡
Avogadro's number	L; N_A	6.0225×10^{23} molecules/mole
Loschmidt's number (molecular density of ideal gas at STP)	L/V_0	2.6870×10^{19} molecules/cm^3
Electron charge	e	4.8030×10^{-10} esu 1.6021×10^{-19} amp sec
Electron mass	m	9.1091×10^{-28} g
Proton mass	m_p	1.6725×10^{-24} g
Ratio of proton mass to electron mass	m_p/m	1836.08
Planck's constant	h	6.6256×10^{-27} erg sec
h-bar	$\hbar = h/2\pi$	1.0545×10^{-27} erg sec
Velocity of light	c	2.997925×10^{10} cm/sec
Faraday's constant	$F = Le$	96,487 amp sec/mole
First Bohr radius	$a_0 = \hbar^2/me^2$	0.5291×10^{-8} cm
Boltzmann's constant	k; k_B	1.3805×10^{-16} erg/degree 8.617×10^{-5} eV/degree
Bohr magneton	$\mu_B = e\hbar/2mc$	9.2732×10^{-21} oersted cm^2
Nuclear magneton	$\mu_N = e\hbar/2m_pc$	5.0505×10^{-24} oersted cm^2
1 angstrom unit	Å	10^{-8} cm
1 micron	μ	10^{-4} cm

‡ After E. R. COHEN and J. W. M. DUMOND, *Rev. Mod. Phys.* **37**, 537 (1965).

TABLE E2.

Conversion Factors of Various Energy Units

	erg	eV	10^{12} Hz	cm^{-1}	degree K
erg	1	6.24179×10^{11}	1.5092×10^{14}	5.0344×10^{15}	7.2437×10^{15}
eV	1.60210×10^{-12}	1	2.41804×10^{2}	8.06573×10^{3}	1.16049×10^{4}
10^{12} Hz	6.625×10^{-15}	4.13557×10^{-3}	1	3.3356×10	4.7993×10
cm^{-1}	1.9863×10^{-16}	1.23981×10^{-4}	2.9979×10^{-2}	1	1.4387
degree K	1.3805×10^{-16}	8.61701×10^{-5}	2.0836×10^{-2}	0.6950	1

Index

OTHER TITLES IN THE SERIES NATURAL PHILOSOPHY

Vol. 40. SOBEL'MAN—Introduction to the Theory of Atomic Spectra
Vol. 41. ARMSTRONG and NICHOLLS—Emission, Absorption and Transfer of Radiation in Heated Atmospheres
Vol. 42. BRUSH—Kinetic Theory, Volume 3
Vol. 43. BOGOLYUBOV—A Method for Studying Model Hamiltonians
Vol. 44. TSYTOVICH—An Introduction to the Theory of Plasma Turbulence
Vol. 45. PATHRIA—Statistical Mechanics
Vol. 46. HAUG—Theoretical Solid State Physics, Volume 2
Vol. 47. NIETO—The Titius–Bode Law of Planetary Distances: Its History and Theory
Vol. 48. WAGNER—Introduction to the Theory of Magnetism
Vol. 49. IRVINE—Nuclear Structure Theory
Vol. 50. STROHMEIER—Variable Stars
Vol. 51. BATTEN—Binary and Multiple Systems of Stars
Vol. 52. ROUSSEAU and MATHIEU—Problems in Optics
Vol. 53. BOWLER—Nuclear Physics
Vol. 54. POMRANING—Radiation Hydrodynamics
Vol. 55. BELINFANTE—A Survey of Hidden-Variables Theories
Vol. 56. SCHEIBE—The Logical Analysis of Quantum Mechanics
Vol. 57. ROBINSON—Macroscopic Electromagnetism
Vol. 58. GOMBÁS and KISDI—Wave Mechanics and its Applications
Vol. 59. KAPLAN and TSYTOVICH—Plasma Astrophysics
Vol. 60. KOVÁCS and ZSOLDOS—Dislocation and Plastic Deformation
Vol. 61. AUVRAY and FOURRIER—Problems in Electronics
Vol. 62. MATHIEU—Optics
Vol. 63. ATWATER—Introduction to General Relativity
Vol. 64. MÜLLER—Quantum Mechanics: A Physical World Picture
Vol. 65. BILENKY—Introduction to Feynman Diagrams
Vol. 66. VODAR and ROMAND—Some Aspects of Vacuum Ultraviolet Radiation Physics
Vol. 67. WILLETT—Gas Lasers: Population Inversion Mechanisms
Vol. 68. AKHIEZER—Plasma Electrodynamics, Volume 1—Linear Theory
Vol. 69. GLASBY—The Nebular Variables
Vol. 70. BIALYNICKI-BIRULA—Quantum Electrodynamics
Vol. 71. KARPMAN—Non-linear Waves in Dispersive Media
Vol. 72. CRACKNELL—Magnetism in Crystalline Materials
Vol. 73. PATHRIA—The Theory of Relativity
Vol. 74. SITENKO and TARTAKOVSKII—Lectures in Nuclear Theory
Vol. 75. BELINFANTE—Measurements and Time Reversal in Objective Quantum Theory
Vol. 76. SOBOLEV—Light Scattering in Planetary Atmospheres
Vol. 77. NOVAKOVIĆ—The Pseudo-spin Method in Magnetism and Ferroelectricity
Vol. 78. NOVOZHILOV—Introduction to Elementary Particle Theory
SRINIVASAN and PARTHASATHY—Some Statistical Applications in X-Ray Crystallography
TAYLOR—Mechanics—Classical and Quantum